T0140375

Multilayer Networks: Analysis and Visualization

Manlio De Domenico

Multilayer Networks: Analysis and Visualization

Introduction to muxViz with R

Springer

Manlio De Domenico
University of Padua
Italy

ISBN 978-3-030-75720-5 ISBN 978-3-030-75718-2 (eBook)
https://doi.org/10.1007/978-3-030-75718-2

This Springer imprint is published by the registered company Springer Nature Switzerland AG
The registered company address is: Gewerbestrasse 11, 6330 Cham, Switzerland

To Serafina, Gabriel and Samuel, for their patience and their love.

Foreword

It's flattering that Manlio De Domenico invited me to write the foreword for this book. After all, the title is "Multilayer Networks: Analysis and Visualization," and that pretty much covers it all you want to know about "Analysis of multilayer networks", and "Visualization of multilayer networks" neatly includes everything else.

Why did I get this task? Partly because I've written articles with Manlio and my scientific respect to him is enormous, and partially because I love this topic and devoted a huge effort to its development.

Perhaps I should say something about what you'll find in this book. It begins, aptly enough, with a selection of concepts about network science and its extension to multilayer networks for beginners. After, the author scrutinizes the structure, dynamics and analysis of multilayer networks with amazing smoothness given the technical roughness of the topics.

The book finishes with the important aspects of visualizing multilayer networks using the tool developed by Manlio, Muxviz. Also some technical notes are included for the computer installation and running of the software, which help the reader to enjoy in first person the visualization power of the tool.

To summarize, I just would like the reader to enjoy the book as much as I did.

Tarragona, July 2021 *Alex Arenas*

Preface

Networks are mathematical objects widely used in multiple disciplines to model the structure of complex systems. From the network of neurons in the human brain to the social networks of humans in our society, *Network Science* quickly gained attention in the last two decades, because of its inherently interdisciplinary approach to modeling and analysis, and the depth of its novel insights.

Network Science is not new: pioneered by social scientists and biologists, its development across half a century is the result of a fruitful cross-fertilization between those disciplines and applied math. Only two decades ago the field attracted the interest of the most visionary physicists with an interdisciplinary mindset, whose contributions led to build what it is known as *Network Science*. Nowadays, this field is a fundamental part of data science and its application spans all domains of knowledge, including physical sciences, life sciences, social sciences and applied sciences. *Network Science* is, among others, a fundamental tool for the analysis of complex systems: the physics of such systems, being them made of particles, molecules, cells or individuals, is one of the most prolific and active research field of the 21st century.

Almost one decade ago, network scientists recognized that the classical methods developed by network scientists were not enough to describe and account for the complexity of a broad spectrum of systems, those ones characterized by multiple types of simultaneous interactions among units and interdependencies. Such systems are better known as *multilayer networks*.

Despite the considerable number of publications and a few volumes dedicated to this novel framework, a comprehensive text concerning the *data science of multilayer networks* is still missing, and this book is a first attempt to fill this gap. My purpose is to provide recipes to perform, in practice, analysis and visualization of empirical multilayer networks, with a wide spectrum of applications, such as in urban transport, human mobility, (computational) social sciences, neuroscience, molecular medicine and digital humanities, to cite the most relevant ones.

As a physicist, I am strongly convinced that our discipline has the potential to play a central role to advance human knowledge. However, as a complex systems scientist, I am strongly convinced that the methodologies developed to study complex systems are revolutionizing our approach to the study of the physical world. This revolution is positively embraced by several disciplines traditionally not related to physics: from systems biology to social sciences, emerging fields like network medicine, econophysics and social physics are the result of intensive interdisciplinary collaborations. Even the most recent advances in the field of computer science, such as artificial intelligence, are starting to benefit from standard concepts developed by physicists. The same holds to the other way around: physics is starting to benefit from other disciplines by employing new methods, such as genetic algorithms or deep learning, to find (quasi–)optimal solutions to standard problems such as the identification of critical points in phase transitions and the characterization of collective phenomena.

Network scientists with different backgrounds, from theoretical physics to applied mathematics, are quickly washing out the traditional borders of knowledge and, in the next future, my hope is that the new generation of physicists will have the opportunity of being formally trained in Complexity Science as they are, nowadays, in Quantum Mechanics or General Relativity.

Since we live in the century of Complexity, as predicted by Stephen Hawking almost two decades ago, and complexity science requires both theoretical and computational skills, it is fundamental to rethink academic books from a novel perspective. The goal of this book is to provide the reader with the basic theoretical foundations behind the computational tool we are going to use, as well as practical guidelines and examples to use them for analyzing the real world. Therefore, the reader interested in a broader theoretical overview about multilayer network science cannot be satisfied only by this work which, instead, should be considered as complementary to textbooks dedicated to that specific purpose.

Accordingly, this book is expected to guide the reader into multilayer network science through developed applications to multiple domains of knowledge. The computational framework adopted for this purpose is the one of **muxViz** : a set of visual tools based on a huge library of functions written in R, developed for the analysis and the visualization of multilayer systems.

The story behind **muxViz** deserves a few lines to be explained, because it perfectly summarizes the interdisciplinary mindset and efforts discussed above. It was the 2013 and the researchers part of the international project PLEX-MATH – among the ones funded within the FP7 FET-Proactive Call 8 Dynamics Multi-level Complex Systems (DyMCS) – were meeting to devise a common research roadmap. At that time, I was a postdoctoral fellow at the Universitat Rovira i Virgili, hosted in the group led by Alex Arenas.

Alex Arenas (Universitat Rovira i Virgili), Marc Barthelemy (CNRS/CEA), James Gleeson (University of Limerick), Yamir Moreno (Universidad de Zaragoza-/BIFI), Mason A. Porter (at that time at Oxford University, now at UCLA), and part of their labs were attending the meeting. One of the many outcomes of that meeting was the call for a user-friendly computational tool to at least visualize multilayer networks and for a library to share with the rest of the academic community. Several proposals were given, but one in particular attracted my attention: the fact that networks could be placed on layers and layers could be arranged in some meaningful way "using some simple projective geometry", as commented by Alex Arenas.

It turned out that while the projective geometry to use was relatively simple, allowing for the flexibility to visualize layers and their networks according to multiple criteria was a more challenging task. The result of almost one year of studies and development was the first version of **muxViz** : a computational framework based on the language R, with a nice graphical user interface. After its release and the publication of its accompanying research paper, **muxViz** quickly (and unexpectedly) became a standard tool for analysis and visualization of multilayer networks, with a fast-growing community of enthusiastic users (more than 600 subscribed to the official group, by the end of December 2020). Nowadays, **muxViz** is free, open source and about 2000x faster than its first version and can count on a library with hundreds of functions for creating, manipulating, modeling, analyzing and visualizing multilayer networks.

Despite the theoretical and computational techniques developed in this new field are uncountable, **muxViz** does not include all of them, mostly because of the lack of publicly available (R) code. Definitively more work is needed in the future, to account for the several algorithms not (yet) included in this framework. In the meanwhile, **muxViz** allows for the generation of multilayer models, the calculation of the most used centrality descriptors, the detection of communities, the reduction of multilayer structures, the analysis of triads and motifs, to mention some features. The future editions of this book will cover novel theoretical and computational tools, from more sophisticated generative models to robustness and percolation analysis: however, note that the current version of the underlying library allows one to write their own R scripts and, in principle, can be suitably used to implement existing algorithm, thus extending the `muxViz` toolkit.

This book is a short journey through the theoretical background of those specific features, from the perspective of an interdisciplinary physicist interested in the analysis of real systems, regardless of their domain. This book requires an open mindset to be read because it is written with an open mindset, where principles of social science are coupled with applications to system medicine or transportation engineering. For a recent work covering more extensively the theoretical aspects of multilayer network science, I refer to the very recent [1] and to [2].

Practitioners and researchers in all disciplines where data allow for a multilayer representation are, in principle, the primary audience for the book. A non-exhaustive list of disciplines includes: physics, neuroscience, molecular and system biology, urban transport and engineering, digital humanities, social and computational social science. The text is accompanied by several code snippets adequately designed to reproduce specific analyses or visualizations, as well as by data sets of real multilayer networks, to facilitate the journey of the reader through the computational aspects of this discipline.

Trento, July 2021, *Manlio De Domenico*

The original version of this book was revised: Affiliation of the author Manlio De Domenico has now been corrected. The correction to this chapter is available at https://doi.org/10.1007/978-3-030-75718-2_8.

Acknowledgements

The writing of this short book has been a long journey allowing me to learn from and interact with many collaborators, outstanding humans and scientists. Of course it is not possible to acknowledge every single person in a short page like this one, therefore I should limit to mention the closest collaborators who helped me to develop the field of multilayer network science (and complexity science in general), with their knowledge and their experience.

For historical reasons, I am happy to start from the Tarragona cluster, with Alex Arenas, Javier Borge-Holthoefer (*ad honorem*), Albert Diaz-Guilera (*ad honorem*), Jordi Duch, Sergio Gòmez, Clara Granell, Roger Guimerà, Joan Matamalas, Elisa Omodei, Marta Sales-Pardo and Albert Solé-Ribalta, to extend my thanks to the Spanish group of network scientists including Alessio Cardillo, Emanuele Cozzo, Jesus Gomez-Gardenes, Sandro Meloni, Yamir Moreno, Jordi Soriano, Sara Teller, and the members of the PlexMath project, including James Gleeson and Marc Barthelemy.

My work on multilayer networks is also due to outstanding collaborations with Jacopo Baggio, Mikko Kivela, Andrea Lancichinetti, Vito Latora, Antonio Lima, Vincenzo Nicosia, Mason Porter, Martin Rosvall, as well as with Eduardo Altmann – during and after my research visit at the Max Planck Institute for Complex Systems in Dresden – and Ivan Bonamassa, Andrea Baronchelli, Jacob Biamonte, Emilio Ferrara, Sandra Gonzalez-Bailon, Shlomo Havlin, Darren Kadis, Marta Gonzalez, Marco Grassia, Giuseppe Mangioni, Shuntaro Sasai, Amitabh Sharma and Arda Halu. I am glad also to acknowledge interesting discussions, during the journey, with Alain Barrat, Dani Bassett, Federico Battiston, Marya Bazzi, Rick Betzel, Marian Boguñá, Ginestra Bianconi, Dirk Brockmann, Guido Caldarelli, Mario Chavez, Aaron Clauset, Vittoria Colizza, Raissa D'Souza, Fabrizio De Vico Fallani, Mattia Frasca, Santo Fortunato, Giuseppe Giordano, Peter Grassberger, Lucas Jeub, Sonia Kefi, Dima Krioukov, Renaud Lambiotte, Dan Larremore, Matteo Magnani, Amos Maritan, Peter J. Mucha, Vera Pancaldi, Pietro Panzarasa, Leto Peel, Tiago Peixoto, Matjaž Perc, Shai Pilosof, Filippo Radicchi, José Ramasco, Mariangeles Serrano, Samir Suweis, Dane Taylor, Alessandro Vespignani, Nina Verstraete, Maria Prosperina Vitale, Brady Williamson and many members of the Complex Systems Society and its Italian Chapter, as well as the Network Science Society.

I am pleased to acknowledge the environment found at the Fondazione Bruno Kessler in Trento, where I have collaborated with Paolo Bosetti, Giuseppe Jurman, Bruno Lepri, Stefano Merler and Piero Poletti on multilayer modeling of biological systems and epidemics processes. Here, the flagship project CHuB (Computational Human Behavior) has allowed me to explore aspects of human behavior by means of statistical physics and big data gathered from online social systems, and develop interesting interdisciplinary collaborations with Pier Luigi Sacco and Sara Tonelli in the field of computational social science. Very special thanks go to the past and the current members of my lab, outstanding young researchers working hard to push forward multilayer network science and its tools (in an epoch where it is too easy to get lost in algorithmic solutions rather than spending efforts to develop grounded measures): Oriol Artime, Barbara Benigni, Giulia Bertagnolli, Sebastiano Bontorin, Nicola Castaldo, Valeria D'Andrea, Riccardo Gallotti, Arsham Ghavasieh, Sebastian Raimondo and Massimo Stella.

The most recent development of `muxViz`, the central actor of this book, has been also possible thanks to the support of a growing community of enthusiastic researchers and practitioners (to date, more than 600) who use, test and help developing the platform. Here, I want to mention the voluntary support from Sergio Alcalá-Corona, Giulia Bertagnolli, Guillermo de Anda Jáuregui, Francesc Font, Rodrigo García, Patricia Gonçalves, Kim Klark, Maria Pereda, Rafael Pereira, Sneha Rajen, Lucio Agostinho Rocha, Hiroki Sayama, Yuriko Yamamoto and Tzu-Chi Yen.

Last, but not least, I feel in debt with my family, Serafina, Gabriel and Samuel, for the time I was able to dedicate to this project thanks to their support and their patience.

Contents

Figures and plots

Boxes

Code Snippets

Acronyms

BA	Barabasi-Albert
CM	Configuration Model
ER	Erdős-Rényi
GCC	GNU C Compiler
HITS	Hyperlink-Induced Topic Search
LFR	Lancichinetti-Fortunato-Radicchi
LIB	Standalone Library
MUX	Multiplex
NIL	Normalized Information Loss
NMI	Normalized Mutual Information
OS	Operating System
PA	Preferential Attachment
PR	PageRank
RW	Random Walk
SBM	Stochastic Block Model
WoK	Web of Knowledge

Part I
Multilayer Network Science: Analysis and Visualization

Chapter 1 | Introduction

Contents

A wide variety of natural and artificial systems are characterized by interactions among their units: from biological molecules within a cell to neurons within the human brain, from individuals in a virtual or physical society to machines in communication networks like the Internet.

Despite their manifest differences, all these systems have a common feature: they are organized into a non-trivial web of relationships which define large-scale structures usually referred to as *complex networks*. In a complex network, entities are represented by nodes (e.g., people) and their interactions are represented by links or edges (e.g., social relationship), as shown in Fig. ***1.1***. Node and links are often enriched with additional information, called metadata, allowing to better characterize the system in terms of different types of entities, their possibly directional interaction or weighted relationships, so forth and so on.

Network Science is that branch of Complexity Science devoted to analyze such large-scale structure in order to gain new insights about system: from the identification of key players (e.g., *influencers*) to the identification of the mesoscale organization into groups or communities. However, Network Science is not limited to the analysis of structural features of a complex system, which is related to the static description of a network. Very often, for practical applications one can be interested into the dynamics of such systems and their response to specific actions or perturbations. A typical example is the analysis of the spreading of epidemic diseases in social systems, where the underlying structure and the human intervention usually play a crucial role for the disease to become endemic or die out within a limited temporal horizon. Similarly, the analysis of how information spreads across different social groups or from the physical layer – the network of social relationships in the physical world: e.g., family, school, business, *etc.* – to the virtual layer – the network of social relationships in the online world: e.g., Twitter, Facebook, WhatsApp, *etc.* – and *viceversa*. In fact, one of the most important fields of research in Network Science concerns the resilience of natural and artificial complex systems to random failures (e.g., airport closure or mobile phone communication cut off due to adverse climate conditions) to targeted perturbations (e.g., terroristic attack to a specific area of urban environments like a city or removal a specific species from an ecosystem) of nodes or links.

The science of networks is very old and its roots can be found in the problem of the seven bridges of Königsberg, solved by the famous mathematician Leonhard Euler in 1736, who developed the foundations of graph theory to demonstrate, rigorously, the absence of a solution. In the successive two centuries, applied mathematicians, economists, ecologists, social scientists and biologists recognized the importance of network modeling in their disciplines, providing valuable contributions to the development of Network Science. Only towards the end of the 20th century this topic attracted the interest of physicists, bringing novel theoretical and computational tools for modeling and analysis of complex networks which led to scientific breakthroughs culminated in the discovery of the mechanisms behind fundamental features such as small-

M. De Domenico, *Multilayer Networks: Analysis and Visualization*, https://doi.org/10.1007/978-3-030-75718-2_1

worldness by Duncan Watts and Steven Strogatz in 1998 [20] and scale-freeness by Albert-Lazlo Barabasi and Reka Albert in 1999 [21].

The scope of this book, however, does not concern with modeling and analysis of such *classical* complex networks, which has been the subject of excellent textbooks (see [22], [23] and [24] for an introduction to structure and dynamics, but also [25–28] for both pedagogical and technical books on this topic) and focused scientific reviews in a variety of applicative domains, such as social sciences [29–31], cognitive and systems neuroscience [32–37, 37–40], systems biology and medicine [41–48], physics [49–61] and ecology [62, 63], just to mention a few.

A thorough review of the most important achievements in Network Science is well beyond the scope of this book and would deserve one or more dedicated books. In this introductory chapter, it is worth mentioning the efforts in explaining how come structural properties observed in empirical networks at intermediate and large scales emerge from simple mechanisms at the microscopic level. Such properties include, for instance, disordered topologies with nodes separated, on average, by distances much shorter than the size of the system [64], a high preference towards triadic closure and fast information exchange [20], heterogeneous connectivity distribution [21, 65] rather than an homogeneous one, organization into hierarchies [66–69] – as observed in of ecological, cellular, technological, and social networks [70] – and groups [71–83] or functional and dynamical modules [13, 84–90]. A crucial role was played by growth models developed to reproduce features observed in social, biological and communication networks [21, 50, 91–97], as well as theoretical tools – mostly inspired or borrowed by statistical physics – for the analysis of complex networks with arbitrary degree distributions [65, 98], often based on the maximization of an entropy measure subjected to specific topological constraints for classical [99–101] and multilevel [102] systems. Least, but not for their relevance to our comprehension of natural and artificial complex systems, the tools for modeling and analysis of system's critical properties which are useful for a broad spectrum of applications, from a better understanding of network robustness to perturbations – from localized static random failures and targeted attacks, to cascade failures – [58, 103–121], to the identification of influential spreaders [122, 123].

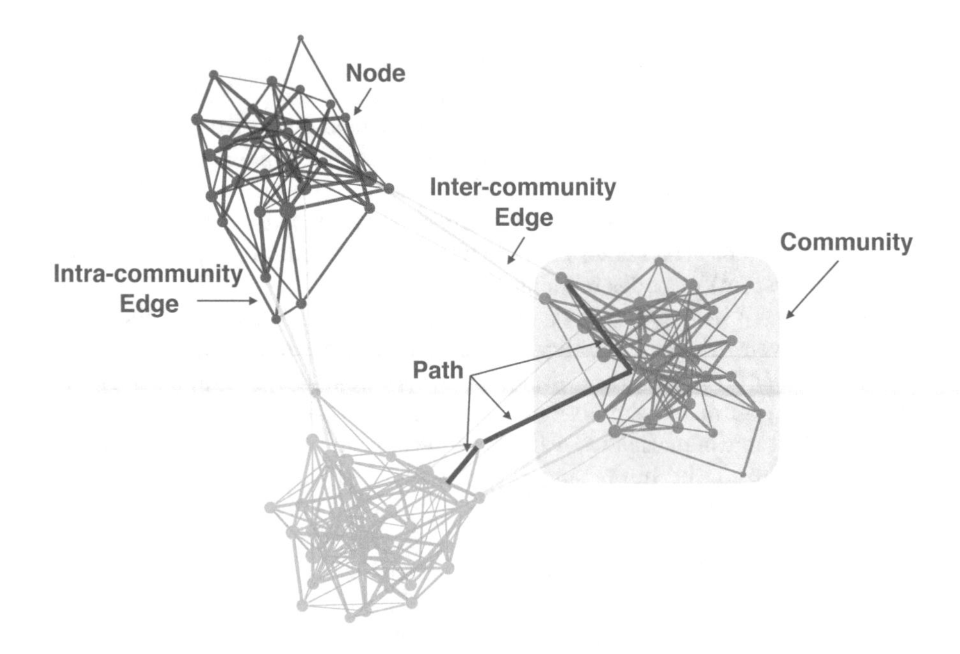

Figure 1.1: Graphical representation of complex network. The system consists of entities, named *nodes* (or *vertices*), connected with each other by *edges* (or *links*) encoding their interactions or their relationships. Usually, nodes tend to cluster into *communities* or *modules*, encoding the modular organization of the system often responsible for its function (e.g., think about teams within a complex company or set of genes and proteins responsible for biological process and cellular components). Nodes which are not connected directly by one edge, can communicate with each other or exchange information through *paths* consisting of intermediate nodes and edges which connect origin and destination. Note, however, that there are network where such an exchange is not possible because nodes are part of disconnected clusters.

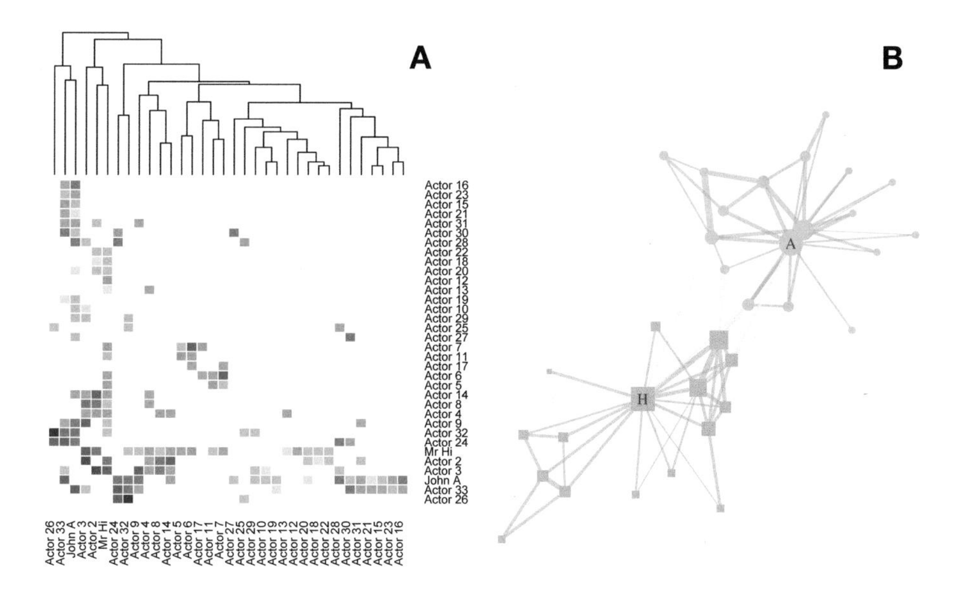

Figure 1.2: Zachary's Karate Club [3] visualized by means of the corresponding weighted adjacency matrix representation (A) and as a network (B). In (A) color codes the weight of social relationships between club members, whose pseudonym are reported. Members are clustered together by applying a complete agglomeration clustering on their Euclidean distance. In (B) color codes the two observed communities, formed after a conflict between Mr. Hi and John A. See the text for further details.

Despite these efforts, some properties of complex biological, social and technological systems continue to elude us. One possible reason might be related to technical difficulties in dealing with either unavailable or noisy, incomplete or multidimensional information. In the first case, a variety of solutions based on Bayesian inference have been proposed [68, 83, 124–127]. In the other case, traditional approaches based on simple or heuristic aggregation of available topological data, as well as approaches neglecting or discarding multidimensional information, often lead to approximate representations of empirical systems, resulting in poor agreement with observation. The core of this book deals with this latest case, covering how multilayer modeling and analysis has been able to cope with theoretical and computational challenges to integrate multiple sources of available data, instead of aggregating them.

A classical example used to show the predictive power of Network Science is Zachary's Karate Club [3]. Wayne Zachary, an anthropologist working at the Department of Anthropology at Temple University in Philadelphia, studied the social relationships among people involved in a University karate club between 1970 and 1972. Zachary built a network model of the group that was able to predict the fission into two smaller groups (see Fig. ***1.2***) as a consequence of a conflict between the two administrators of the club (anonymized as Mr. Hi and John A). The study, published in 1977, released the network data (see Fig. ***1.2***) which became a standard benchmark for community detection algorithms in Network Science[1] and a special prize is assigned to network scientists who, during their oral presentation at conferences in the field, first mention this data set[2].

Zachary's analysis of the network made use of the Ford-Fulkerson method [128], maximizing flow in a graph where links can be interpreted as information channels and their weights are proportional to the underlying flow. Using this method on the original network, Zachary was able to determine – with the exception of one member – the two subgroups formed after the scission.

However, Zachary had to face a complex data analysis problem. In fact, he identified eight different "contexts" in which he observed and measured social relationships, from association in and between academic classes at the University to interactions in different locations or attendance at Karate tournaments. Without theoretical and computational tools at hand to deal with such an amount of information, he investigated the possibility to aggregate the data. In practice, for each pair of members, Zachary was able to build an 8-dimensional vector encoding their relationships (0 if absent, 1 otherwise) across different contexts but, due to the lack of an adequate mathematical framework, he even-

[1] A joke circulating among network scientists reports that "*if your method does not work on this network, then go home*".

[2] http://networkkarate.tumblr.com/

tually summed up the entries, linearly. Despite his attempts to provide arguments to sustain his claim about linearity of data aggregation, 40 years later we would assess that he had to deal with an edge-colored multigraph – a special type of multilayer network – and he decided to analyze its aggregate network representation.

1.1 Mathematical representation of a complex network

From a mathematical point of view, a convenient way to represent a complex network is to encode the underlying information about adjacent nodes into a square matrix. If we indicate by $\mathbf{W}$ the *adjacency matrix* of a network with N nodes, the entry w_{ij} is usually a real positive number if there is a direct connection from node i to node j $(i, j = 1, 2, ..., N)$, otherwise $w_{ij} = 0$.

The fact that complex networks can be represented by matrices suggested the possibility to exploit classical linear algebra for their analysis. More recently, a rigorous mathematical formulation in terms of rank–2 tensors has been introduced [129], which turned out to be extremely useful for generalizing the algebra classical of complex networks, also named *monoplex* networks, to multilayer networks, the subject of this book. In this formalism, the adjacency matrix $\mathbf{W}$ is nothing but a rank–2 tensor W^i_j, where i and j are covariant and contravariant indices encoding the two dimensions of the adjacency tensor.

The general reader, specially the one less familiar with linear algebra, might underestimate the importance of using a tensorial representation for complex networks. For sake of completeness, we briefly provide further details in Box 1.1.1 and Box 1.1.2 as an advanced topic.

Box 1.1.1: What is a tensor?

A tensor is a multilinear mathematical object that, independently on the choice of the underlying coordinate system, maps tensors in other tensors. Scalars (a), vectors (a_i) and matrices (A^i_j) might be recognized as the simplest tensors of rank 0, 1 and 2 respectively. It is easy to capture the pattern: the total rank is given by the number of tensorial indices. In general, tensors are characterized by multiple indices representing two different types of coordinates, usually named covariant (bottom) and contravariant (top), which transform differently under a change of basis. This transformation can be defined by observing that, if the change of basis is governed by a tensor Q, then the contravariant indices must change with Q while covariant indices must change with its inverse, Q^{-1}. Therefore, a rank–1 tensor can be either a 1–covariant (a_i) or a 1–contravariant (a^i) vector, whereas an object like A^{ij}_{klm} is a rank–5 tensor, 2–contravariant and 3–covariant.
To follow the mathematical content of this book, it is sufficient to understand what a tensor is, why it is important and how to work with it. To this aim, we present a few basic operations which are widely used in the reminder of this book.
The **outer product** – also known as the Kronecker product – between two tensors A and B is a new composite tensor C with higher rank. The rank of C is equal to the sum of the two ranks of A and B. In fact, this property holds for the number of covariant and contravariant indices. For instance: $A^i_{jk}B^l_m = C^{il}_{jkm}$.
Another common operation is the **inner product**, also known as **contraction**, where the resulting tensor has a rank which is equal to the sum of the two ranks reduced by 2. For instance: $A^i_{jk}B^k_m = C^i_{jm}$. If there are multiple contractions, the total rank will be further reduced: $A^i_{jk}B^j_i = C_k$. Here, we have implicitly adopted the Einstein summation convention to repeated indexes, in order to reduce the notational complexity of our tensorial equations. In fact, the full notation for contractions is as follows:

$$A^i_i = \sum_{i=1}^{N} A^i_i, \qquad A^i_j B^j_i = \sum_{i=1}^{N}\sum_{j=1}^{N} A^i_j B^j_i$$

whose right-hand sides include the summation signs explicitly and N is the size of the space where the two tensors are defined. When working with tensors, it is crucial to specify it, because in another notation A_{ij} might simply indicate the scalar entry of the matrix $\mathbf{A}$ corresponding to row i and column j, whereas in tensorial notation this object would represent a rank–2 tensor.

Box 1.1.2: The adjacency tensor of a complex network

The adjacency tensor W^i_j of a network can be represented [129] as a linear combination of tensors of the canonical basis by

$$W^i_j = \sum_{a,b=1}^{N} w_{ab} e^i(a) e_j(b) = \sum_{a,b=1}^{N} w_{ab} E^i_j(ab), \quad (1.1)$$

where w_{ab} encode the weight of the interaction between nodes a and b, $E^i_j(ab) \in \mathbb{R}^{N\times N}$ indicates the tensor of the canonical basis corresponding to the tensorial product of the canonical vectors $\mathbf{e}(a) \equiv e^i(a)$ and $\mathbf{e}^\dagger(b) \equiv e_j(b)$, column and row respectively, both defined in $\mathbb{R}^N$, assigned to nodes a and b, respectively. Here, the natural basis is the Euclidean one.
The assignment of the indices as covariant or contravariant may seem arbitrary because of the lack of a natural transformation that can be used to guide us. For monoplex networks, the adjacency tensor W^i_j is nothing but a linear transformation which returns the set of nodes adjacent to a specific one (e.g., node i) when applied to a rank–1 tensor (i.e., a vector) representing that node: $W^i_j e_i(a) = w_j(a)$. Since complex networks can be directed, the necessity to distinguish between adjacent nodes with incoming links and adjacent nodes with outgoing links makes 1–covariant 1–contravariant tensors the only possible choice.
The knowledge of the order of the adjacency tensor completely determines its transformation under a change of coordinates. In fact, let

$$Q^i_j = \sum_{a=1}^{N} e'^i(a) e_j(a) \quad (1.2)$$

be the change of basis tensor which transforms the basis vector set $\{e^i(a)\}$ into a second set $\{e'^i(a)\}$. Since any change of basis should not alter w_{ab} we obtain:

$$\begin{aligned} W'^k_l &= \sum_{a,b=1}^{N} w_{ab} e'^k(a) e'_l(b) = \sum_{a,b=1}^{N} w_{ab} Q^k_i e^i(a) e_j(b) (Q^{-1})^j_l \\ &= Q^k_i \left[\sum_{a,b=1}^{N} w_{ab} e^i(a) e_j(b) \right] (Q^{-1})^j_l = Q^k_i W^i_j (Q^{-1})^j_l, \end{aligned} \quad (1.3)$$

demonstrating the tensorial nature of W^i_j. It is worth remarking that the existence of such a transformation law makes the adjacency tensor an object much richer than the adjacency matrix, although they can be both represented as an array of arrays, or equivalently an hypermatrix of order 2. While the components of a tensor can be arranged into hypermatrices, the components of hypermatrices do not necessarily define a tensor.

1.2 Multilayer networks: towards a more realistic model of complex systems

By the end of the '60s a few visionary biologists, sociologists, psychologists, physicists and mathematicians were already convinced that interacting units, from molecules in the cell to individuals in a society, build an interconnected network whose emergent behavior – characterized by spontaneously appearing phenomena – can not be understood from the analysis of its components in isolation[3].

[3] The reader interested in the development of system thinking in the past century might want to read the book by Capra and Luisi [130].

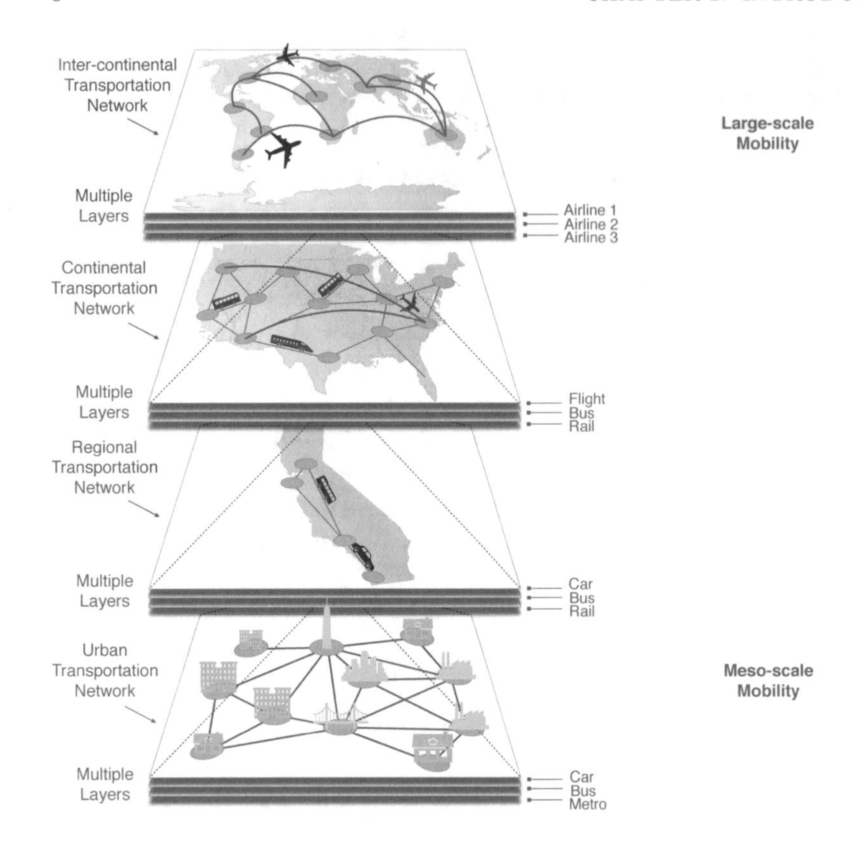

Figure 1.3: A schematic representation of a multilayer transportation system across several geographic scales. From top to bottom: large-scale mobility network encoded by distinct airlines (the layers) serving inter-continental routes; continental transportation (here in the USA) with respect to domestic flights, long-range bus routes and rail (the layers); at a lower scale, e.g., in California, one can use the road network, as well as bus routes and rail (the layers) for regional movements; at the lowest scale one can use roads, buses and metros (the layers) to move between distinct parts of a city. Figure from [4] under Creative Commons Attribution-ShareAlike 4.0 International License.

Sociologists and biologists, however, were among the first, if not the first at all, to recognize that natural and social networks are often organized in structures more complex than classical networks. While multilevel organization characterizes living organisms and complex societies, other fundamental ingredients of complexity are multiplexity and interdependency between different systems. When observed from a wider perspective, Nature seems to be favor *systems of systems*: networks consisting of other networks characterized by non-trivial, and often heterogeneous, interdependencies or simultaneous relationships that can not be either simply neglected or aggregated. Molecules interact in different ways within cells, cells combine to form tissues, tissues interact to form organs and organs are complex systems interdependent from each other forming organisms. Remarkably, organisms interact in different ways to build multiplex and interdependent social systems and ecosystems.

In 1969, Brian Kapferer defined multiplexity as social engagement in different types of exchanges [131], and Mark Granovetter, a few years later, argued that "*the degree of overlap of two individuals' friendship networks varies directly with the strength of their tie to one another*". Social scientists empirically demonstrated how multiplexity can have a direct impact on how influence and information are diffused through the system, as well as on how individuals organize into communities [132]. Almost 40 years later, physicists and applied mathematicians were fascinated by the same challenges, developing a mathematical theory of multilayer networks that confirmed, analytically, those intuitions about diffusive processes [18, 129, 133] and group organization [15, 87]. Lois Verbrugge identified that multiplexity occurs when actors share multiple bases for interaction in a dyad [134], while John Padgett demonstrated the role of multiplexity in the rise of the Medici among families in the Renaissance Florence [135, 136].

On the one hand, social scientists were among the first to recognize the role of multiplexity in shaping complex societies. On the other hand, system biologists were among the first to recognize the role of interdependency among

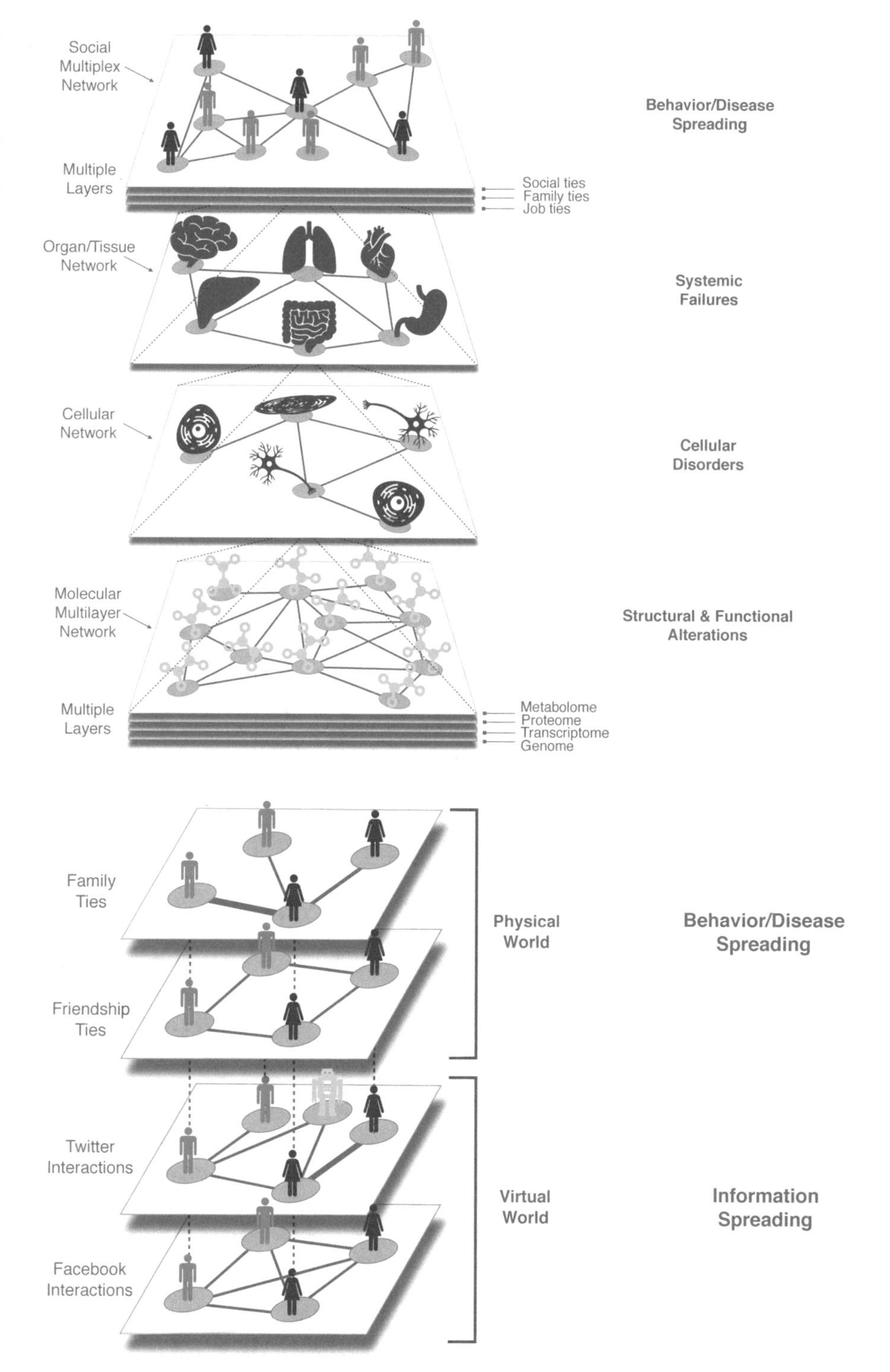

Figure 1.4: A schematic representation of a multilayer biological system across several biological scales. From top to bottom: a social network of individuals where layers characterize distinct ties (e.g., social, family, work); each individual is an organism characterized by a network of organs which characterize distinct systems (e.g., circulatory, respiratory, nervous, so forth and so on); tissues are characterized by cellular networks and, at the bottom, each cell is characterized by molecular interactions of distinct type which constitute the genome, the transcriptome, the proteome and the metabolome, for instance. Figure from [4] under Creative Commons Attribution-ShareAlike 4.0 International License.

Figure 1.5: A schematic representation of a multilayer social system. Layers here encode distinct type of interactions and actors: physical individuals interact in the physical world with different ties (e.g., family and friendship), whereas they can have digital counterparts – such as accounts on social media platforms – with other relationships (e.g., following on Twitter, friendship on Facebook). In the digital layers we can see also the presence of non-physical actors, such as social bots [5, 6]. This type of structures allows one to model complex population dynamics, such as epidemics and behavioral spreading as well as information diffusion. Figure from [4] under Creative Commons Attribution-ShareAlike 4.0 International License.

different systems, from cells to organs within an organism, as remarked in the famous sentence of Francois Jacob – Nobel Prize in Physiology or Medicine in 1965 –: "*Every object that biology studies is a system of systems.*" [137].

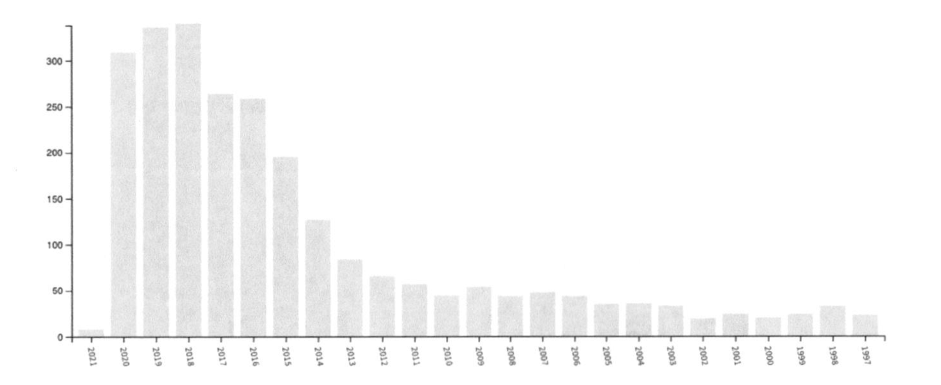

Figure 1.6: Bibliometrics analysis from the Web of Knowledge (WoK). Temporal evolution (from right to left) of the number of papers published about multilayer networks, including more than 2600 publications between 1990 and 2021 (2100+ since 2009 to date). Obtained by filtering with the words "multiplex network", "multilayer network" and "interdependent network" in the title/abstract, while requiring the word "complex" to be within 10 words and excluding keywords used in other contexts. Data and visualization from webofknowledge.com.

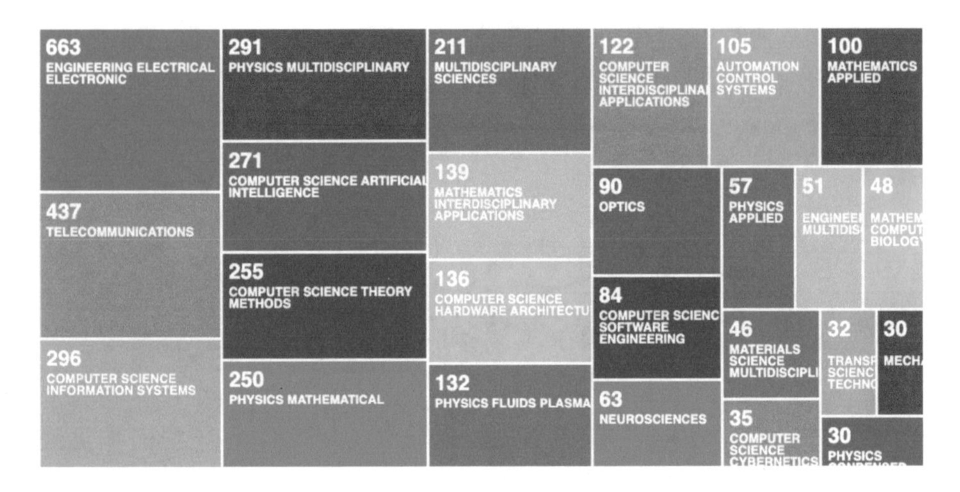

Figure 1.7: Same data used in Fig. ***1.6*** to show the WoK categories where papers on multilayer networks have been mostly published. Data and visualization from webofknowledge.com.

Despite its old roots, the science of multilayer networks flourished only one decade ago, when it became clear that the robustness of a complex system is strictly related to its interdependencies, with potentially catastrophic consequences[4] triggered by their existence [139, 140]. In parallel, the role of multiplexity[5] in determining critical functional units in social systems was mathematically and computationally formalized [87], defining *de facto* the new era for multilayer network science.

In the last decade, multilayer network science [141–145] provided great opportunities for more realistic models and meaningful analyses of empirical systems, being one of the best candidates for a science of integrated systems, especially biological ones [146, 147]. Figures ***1.3***, ***1.4*** and ***1.5*** show illustrative examples of how one can use multilayer networks to model a broad class of complex systems, including transportation, biological and social ones, respectively. Note that such illustrative examples have used to analyze real-world systems in thousands of scientific papers during the last decade, as show in Fig. ***1.6***, ***1.7***, ***1.8***, and ***1.9***, highlighting the interdisciplinary and multidisciplinary nature of multilayer network modeling[6].

In the first part of this book we will explore some important tools and concepts developed to account for multiple types of relationships between nodes, simultaneously. In the remaining parts we will use those tools to analyze empirical systems from a broad spectrum of disciplines and applications.

[4] An emblematic example is given by the chain of events triggered by, and contributing to, climate change – Earth's surface air temperature increase, intensification of its hydrological cycle and risk of river floods, so forth and so on – that will dramatically effect the global trade network and, consequently, will cause relevant economic losses to countries such as China and United States [138].

[5] Mathematically, systems changing over time, also known as *time-varying networks* [56, 57], can be described within the multilayer framework [87, 129], although more research is needed to establish a more formal bridge.

[6] For reproducibility, this is the search query: (((multiplex NEAR/3 network*) OR (multilayer NEAR/3 network*) OR (interdependent NEAR/3 network*) NEAR/10 complex) NOT (feedforward network OR feed-forward network OR neural OR adversarial OR electronic* OR wavelength OR optical OR (internet of things) OR microgrid OR radio OR iron OR supervise* OR wireless))

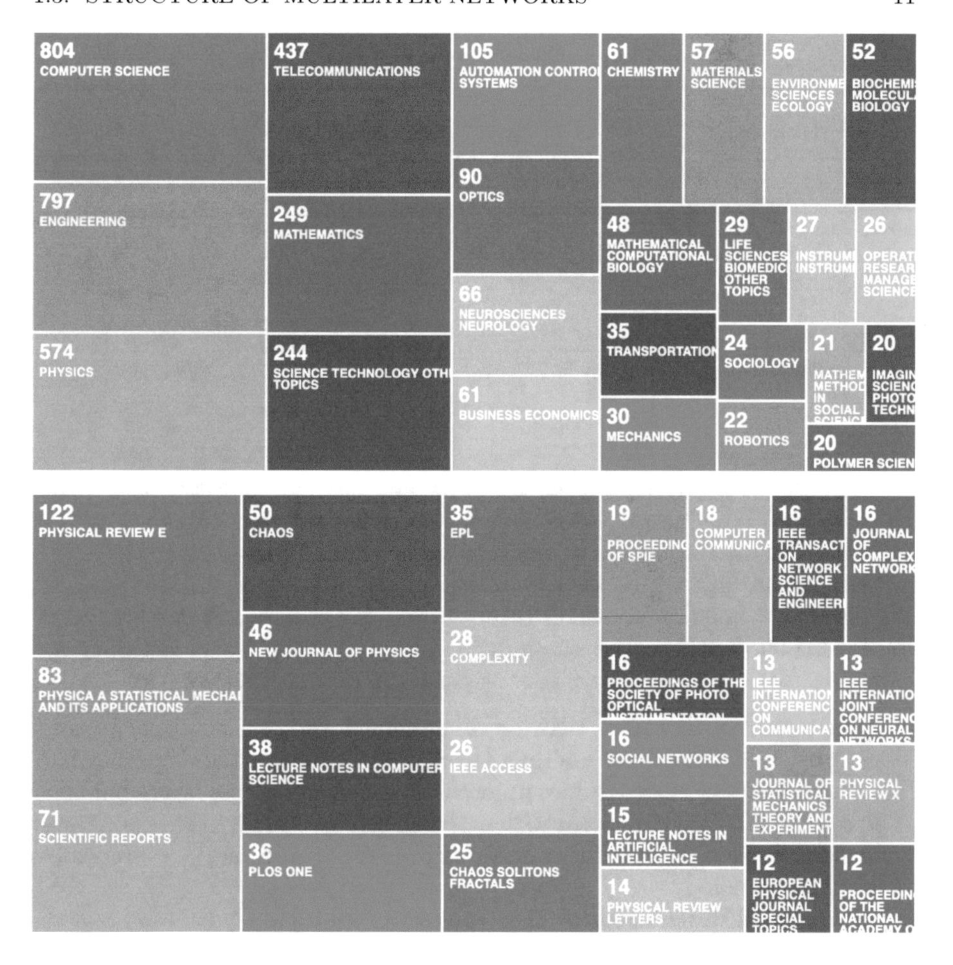

Figure 1.8: Same data used in Fig. ***1.6*** to show the research domains where papers on multilayer networks have been mostly published. Data and visualization from webofknowledge.com.

Figure 1.9: Same data used in Fig. ***1.6*** to show the scientific journals where papers on multilayer networks have been mostly published. Data and visualization from webofknowledge.com.

1.3 Structure of multilayer networks

At variance with monoplex systems, where only one type of relationship among nodes is allowed, a multilayer network is defined by a set of nodes interacting with each other in multiple ways, simultaneously. Each type of relationship is encoded by a "color" and the set of all interactions of the same color defines a *layer*, as shown in Fig. ***1.10***. The same node(s) can exist in one or multiple layers: when the layer information is retained the node is a *state node* or also *replica node*, whereas when we refer to a node regardless of which layer it be-

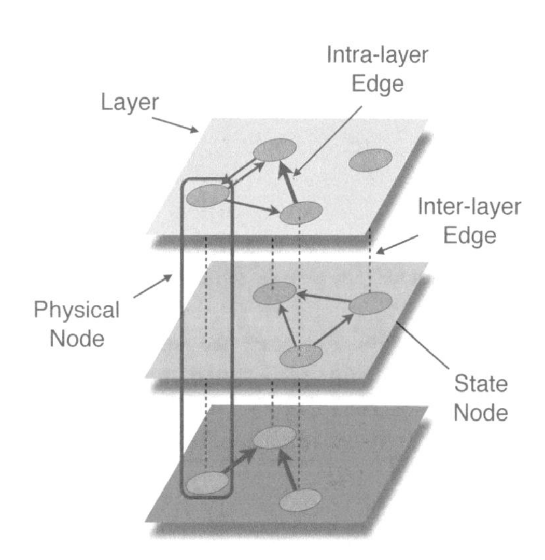

Figure 1.10: Illustration of a typical multilayer network. Layers encode different types of relationships among nodes, occurring simultaneously. Nodes can exist in multiple layers, but not necessarily in all layers: if their identity is related to the a specific layer they are named state nodes, whereas physical nodes are the ones whose identity does not depend on any specific layer. Connectivity within each layer is defined by intra-layer edges, whereas inter-layer edges define connections across layers. Figure from [4] under Creative Commons Attribution-ShareAlike 4.0 International License.

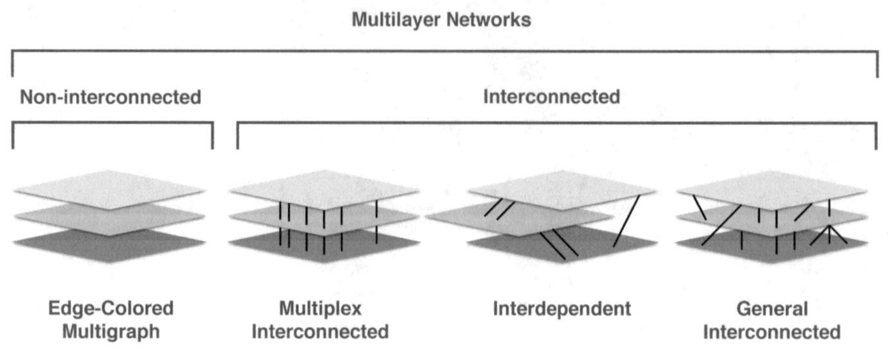

Figure 1.11: Basic classification of multilayer network models. All models can be analyzed and visualized within the `muxViz` platform (see Sec. 1.5). Allowed models include both non-interconnected – i.e., edge-colored – and interconnected – i.e., multiplex, interdependent and general multilayer – networks. Figure from [4] under Creative Commons Attribution-ShareAlike 4.0 International License.

longs to, we use the term *physical node*. Edges of the same color, the ones within the same layer, define the *intra-layer* connectivity, whereas edges connecting state nodes across layers define the *inter-layer* connectivity.

Despite the exponential increase of scientific studies on multilayer systems, the efforts for building a general theory of multilayer networks are still ongoing [129, 143]. The simplest classification of multilayer systems identifies two categories based on the absence or the presence of inter-layer connectivity:

- **Non-interconnected networks**: often named also **edge-colored multigraphs**, they consists of multiple layers, each one encoding a specific relationship between nodes. Nodes preserve their identity across layers but their states are not interconnected with each other (Fig. ***1.11***). State nodes exist in at least one layer and their relationships in different layers can be encoded by different colors.
- **Interconnected networks**: they consists of multiple layers, each one encoding a specific relationship between nodes. Nodes can preserve their identity across layers and their states can be interconnected with each other (Fig. ***1.11***).
  - **Multiplex interconnected networks**: only inter-layer connections among states of the same physical nodes are allowed. In practice, this corresponds to the case of an edge-colored multigraph with interconnected layers.
  - **Interdependent networks**: only inter-layer connections among states of different physical nodes are allowed.
  - **General interconnected network**: no restrictions on inter-layer connections are imposed.

Across this book we will introduce appropriate multilayer models for empirical networks found in different disciplines, such as Social Sciences, Digital Humanities, Engineering, Biology and Biomedicine.

The fine classification of multilayer networks allows for some flexibility in their representation and data format (see Sec. A.5). However, it is worth noting that standard matrices and rank–2 tensors, used to represent monoplex networks, are inherently limited in the complexity of the relationships that they can capture, i.e., they do not represent a suitable framework in the case of multilayer systems, either they are interconnected or not. For instance, this is the case of increasingly complicated types of relationships —that can also change in time— between nodes.

The less complex multilayer systems are edge-colored multigraphs, also known as non-interconnected multiplex networks, which can be represented by an array of adjacency matrices [148–153]. The corresponding systems can be mathematically represented by hypermatrices of order 3 and, under some restrictions, by rank–3 tensors.

When the structure of relationship is very rich, i.e. inter-layer interactions between pairs of nodes are allowed, a more general model is required to encode

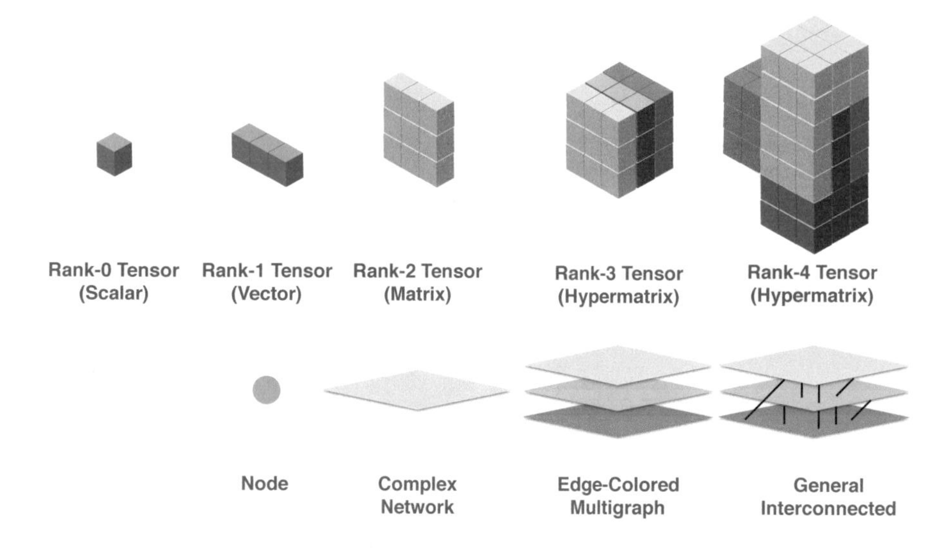

Figure 1.12: Schematic illustration of how tensorial objects – such as scalar, vector, matrix and hypermatrix – of increasing order can be used to encode information about physical objects with increasing complexity, e.g. a node, a network, a non-interconnected multiplex and a general interconnected multilayer system. Figure from [4] under Creative Commons Attribution-ShareAlike 4.0 International License.

all the available information. Intuitively, one might guess that to account for all possible relationships between a node i in a layer α with a node j in layer β, a tensorial object with at least 4 indices is required. In fact, it can be shown that rank–4 tensors, that we will indicate with the notation $M^{i\alpha}_{j\beta}$, are enough for this purpose (see Box 2.3.1 for details). To avoid confusion, in the following we will use Latin letters to indicate nodes and Greek letters to refer to layers.

The object $M^{i\alpha}_{j\beta}$ is usually referred to as *multilayer adjacency tensor* and it accounts for all interconnections among nodes within layers and across layers. To better understand this complicated object, it is useful to inspect Fig. ***1.12***, where an illustration of the "shape" of each tensor corresponding to a specific object, from a node to a multilayer network, is shown. While non-interconnected multiplex systems can be represented by a three-dimensional array, the geometry corresponding to a multilayer network is inherently four-dimensional. However, it is possible cope with the complexity of this object by flattening it to a lower-dimensional object, a rank–2 tensor, by means of the operation known as *matricization* [7]. If there are N nodes and L layers in the system, the multilayer adjacency tensor is defined in a space with $N \times N \times L \times L$ dimensions while its flattening, referred to as *supra-adjacency matrix* [133], is defined in a space with $NL \times NL$ dimensions (see Fig. ***1.13***). While the overall number of dimension is the same, remarking that the information content of the two object is the same, the two underlying spaces are clearly different: the advantage of the second one is that it allows us to work with standard matrix theory provided that some care is taken when results are obtained and must be interpreted. In fact, the rank–2 *intra-layer adjacency tensors* denoted by $C^i_j(\alpha\alpha) = W^i_j(\alpha)$ ($\alpha = 1, 2, ..., L$), with dimension $N \times N$, represent the single-layer networks of the system and are placed, conventionally, on the diagonal blocks of the supra-adjacency matrix. The rank–2 *inter-layer adjacency tensors* denoted by $C^i_j(\alpha\beta)$ ($\alpha, \beta = 1, 2, ..., L$ and $\alpha \neq \beta$), with dimension $N \times N$, represent the interactions between pairs of nodes across layers and are placed, accordingly, on the off-diagonal blocks of the supra-adjacency matrix. The careful reader recognizes that we have defined a total of L^2 rank–2 tensors with dimension $N \times N$.

In practice, the main advantage of the multilayer adjacency tensor is that it encodes all the information in a single object where the identity of each node is uniquely defined by the rank–4 tensor, while this is no more the case when the supra-adjacency representation is used. In fact, in the second case, we must be careful to the (arbitrary) order in which adjacency matrices of single layers are placed in the diagonal block and, overall, it must be remarked that if standard network algorithms are employed they will interpret the supra-adjacency matrix as the adjacency matrix of a network with $N \times L$ nodes, sometimes known also

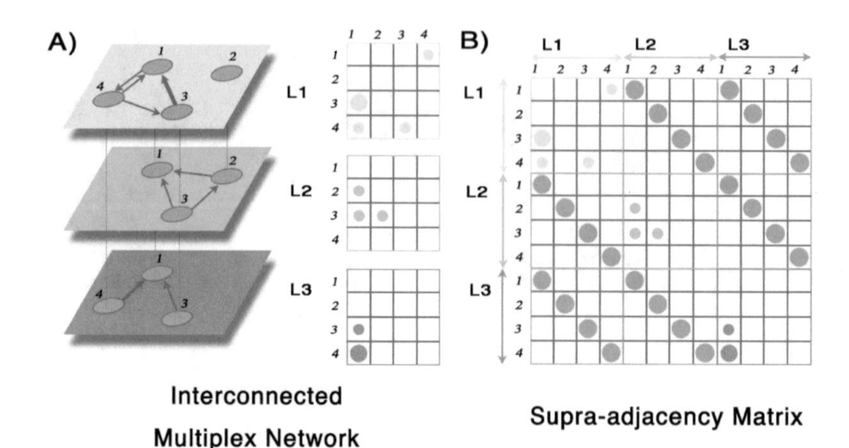

Figure 1.13: Representing multilayer networks with supra-adjacency matrices. (A) An interconnected multiplex network (left) with $N = 4$ nodes and $L = 3$ layers (note that not all nodes necessarily exist on all layers). Each layer is encoded by a different color. The network is directed (encoded by arrows) and weighted (encoded by edge thickness). The adjacency matrices, corresponding to each layer separately, are also shown (right). (B) Matricization [7] is applied to flatten the rank–4 tensor representing the multilayer network to a rank–2 supra-adjacency matrix – a block matrix, i.e., a matrix consisting of matrices of lower dimension – in which: i) the adjacency matrices corresponding to each layer are placed as blocks on the main diagonal, and ii) inter-layer connectivity is encoded into diagonal matrices that are placed on the off-diagonal blocks. This representation preserves the topological information, although some care is necessary to deal with analysis (see the text for details). Reproduced with permission from Ref. [8].

as the *expanded representation* [15]. However, in this expanded representation the identity of physical nodes is demultiplexed into the different identities of the corresponding replica nodes. To cope with this problem of interpretability, it is possible to demonstrate that a wide spectrum of algorithms and methods can be still used safely provided that results are combined appropriately by using tensorial algebra [10, 129] (see Chap. 2–6 for details).

This new framework allows for a natural integration of multiple sources of information, opening the doors for countless solutions and applications in several disciplines, especially in biology [146, 147].

For instance, applications to ecology have been recently considered [9, 154–156]. The way animals are related (e.g., genetically) or interact (e.g., mating) with each other can influence and is influenced by their habitat and its organization in spatial patches (Fig. ***1.14***). Accounting for such a socio-spatial interdependence has the potential to provide new insights on animal behavior and the organization of ecological systems (Fig. ***1.15***).

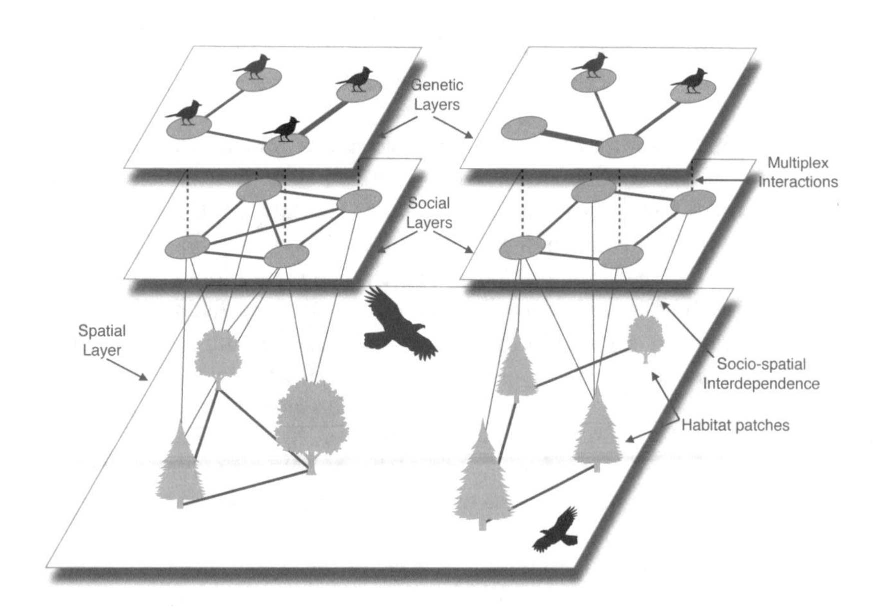

Figure 1.14: A multilayer network model to integrate social, spatial and behavioral information from animals and their environment. Interactions and relationships between animals can be modeled by multiplex networks which, in turn, are interdependent with a spatial network of habitat patches. Figure from [4] under Creative Commons Attribution-ShareAlike 4.0 International License.

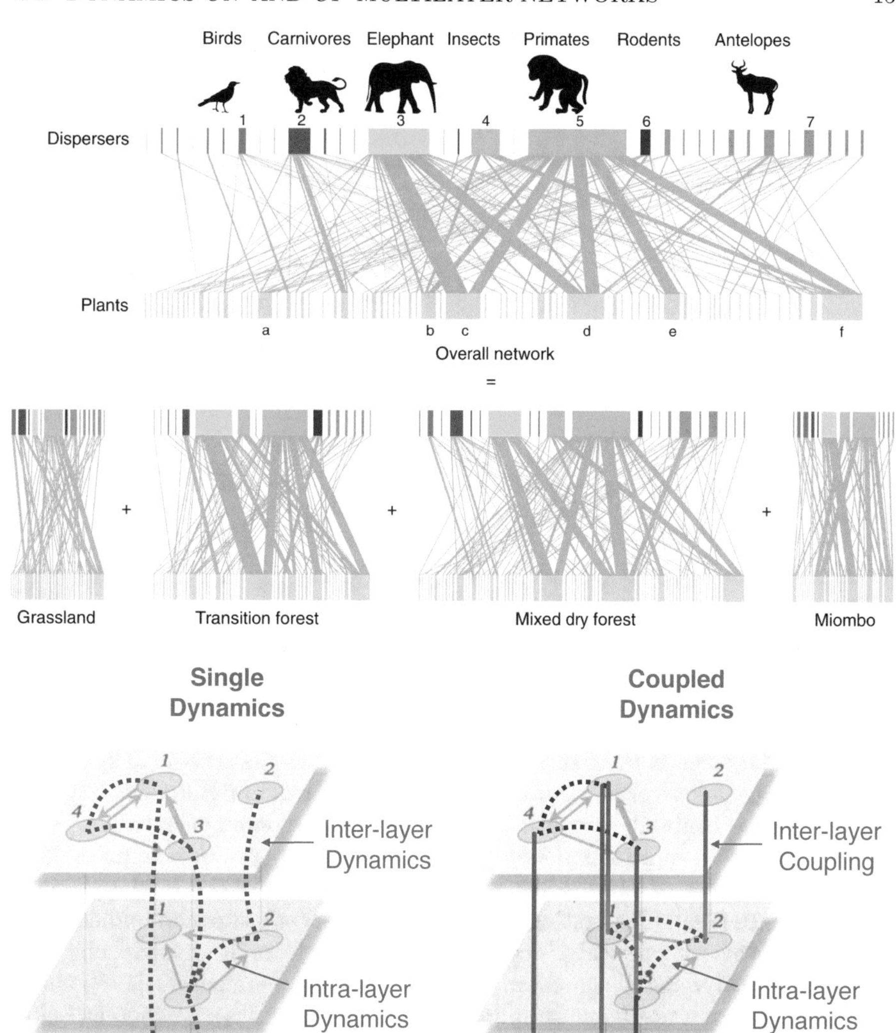

Figure 1.15: Aggregate (top) and multilayer (bottom) representations of a real seed-dispersal network. Reproduced from Ref. [9] under Creative Commons Attribution 4.0 International License http://creativecommons.org/licenses/by/4.0/

Figure 1.16: Types of dynamics on the top of multilayer structures. In single dynamics (*left*) a single dynamical process (dotted line) takes place on the top of the while multilayer topology. In coupled dynamics (*right*), a different dynamical process (dotted lines with different colors) takes place on the top of each layer separately, being intertwined by the underlying inter-layer connectivity (vertical, gradient-colored lines). Figure from [4] under Creative Commons Attribution-ShareAlike 4.0 International License.

1.4 Dynamics *on* and *of* multilayer networks

We have seen so far that the multilayer framework allows a natural representation of interdependent networks in terms of coupled topologies. In fact, a similar feature characterizes dynamical processes that, in a multilayer framework, can be categorized in two distinct classes [145] characterized by either i) single dynamics or ii) coupled dynamics *on* the top of the structure (Fig. ***1.16***). Both classes reveal a variety of interesting phenomena [157], such as structural and dynamical phase transitions, as well as the emergence of enhanced diffusion or congestion effects with respect to the case where layers are considered separately, which uniquely characterize multilayer systems (we refer to [142–145] for a review).

The first class defines dynamical processes such as continuous [133, 158–161], and discrete [18, 162–164] diffusion, as well as synchronization dynamics [165–169], system control [170, 171], cooperation [172–174], communicable information [175], epidemics spreading of a single disease [176–180], traffic and congestion in multimodal communication systems [181–183], opinion dynamics [184–187], innovation diffusion, adoption, simple or complex contagion [188–

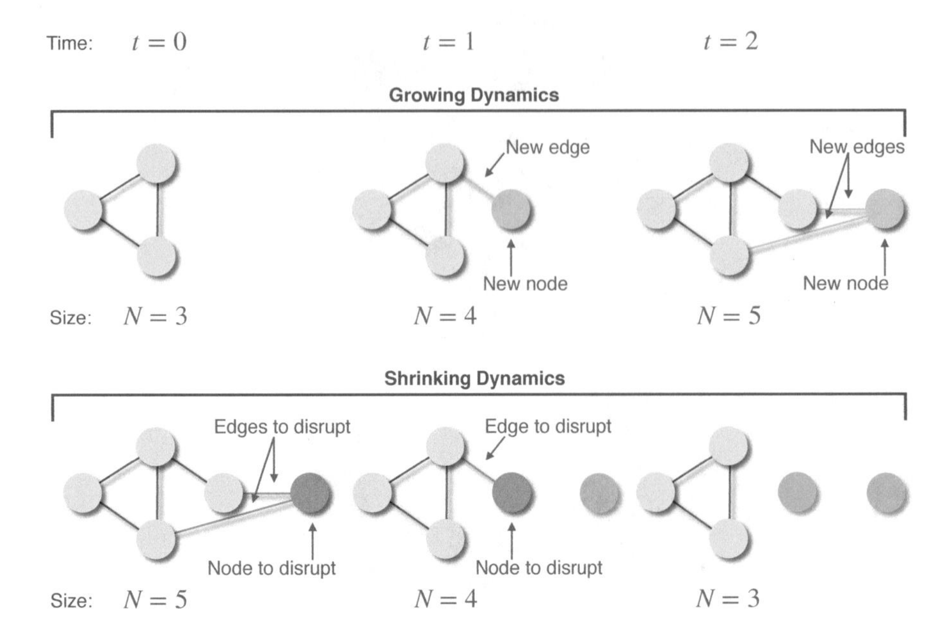

Figure 1.17: Types of dynamics of classical, single-layer networks. The network can grow over time (*top*) by adding new nodes and edges, or it can shrink (*bottom*) because of node and edge failures. Hybrid dynamics, where the two processes are mixed, can be defined to model the dynamics of complex adaptive systems such as social networks, where individuals join and leave groups, establish new social relationships or cut existing ones. Figure from [4] under Creative Commons Attribution-ShareAlike 4.0 International License.

192] such as in viral diffusion of a piece of information (e.g., a meme) or of a product. The emergence of such effects in diffusive processes and in spreading process that, at first order, can be described by diffusion dynamics, is mostly related to the coupling strength between layers. In some special cases, such as interconnected multiplex networks with identical coupling, it is possible to show the existence of two distinct regimes as a function of the inter-layer coupling strength [193], highlighting how the multilayer structure can influence the outcome of several physical processes. Multilayer effects are usually observed above a critical value which characterizes a structural transition [193], corresponding to a non-negligible coupling strength between layers, whereas below such a value, the networks corresponding to different layers tend to act in isolation and can be studied separately. The importance of this type of dynamics for the analysis of multilayer networks will become even more clear in Chap. 2, where we will show how a variety of structural indicators, from node centrality to organization in communities or clusters, can be defined or better understood in terms of specific diffusive processes, such as random walks.

The second class defines the dynamics of different processes – in which each one runs on top of a given layer – which are eventually coupled together by the underlying multilayer topology. The existence of the topological coupling between layers and its strength are responsible for emerging phenomena, of interest for applications in molecular biology, neuroscience, economics, engineering and social sciences. These phenomena depends on interdependent dynamics, such as combining cooperative and/or competitive epidemics spreading [194–198], interplay between epidemics spreading and behavior [199–208], simple and complex contagion [209], evolutionary game dynamics and social influence [210], coupled human mobility [211], transport and synchronization dynamics [212], as well as other collective phenomena [213]. Despite the different contexts, in all cases the two dynamics can have positive or negative feedbacks: for instance, behavior (e.g., information awareness) can inhibit the disease spreading; social mixing between classes and mobility can produce abrupt changes of the critical properties of the epidemic onset; cooperation emerges when classical expectation was defection. This leads to the interdependence between the corresponding critical points of the dynamics characterized by the existence of a curve of critical points separating two different regimes: i) one in which the critical properties of one process do not depend on those of the other, and (ii)

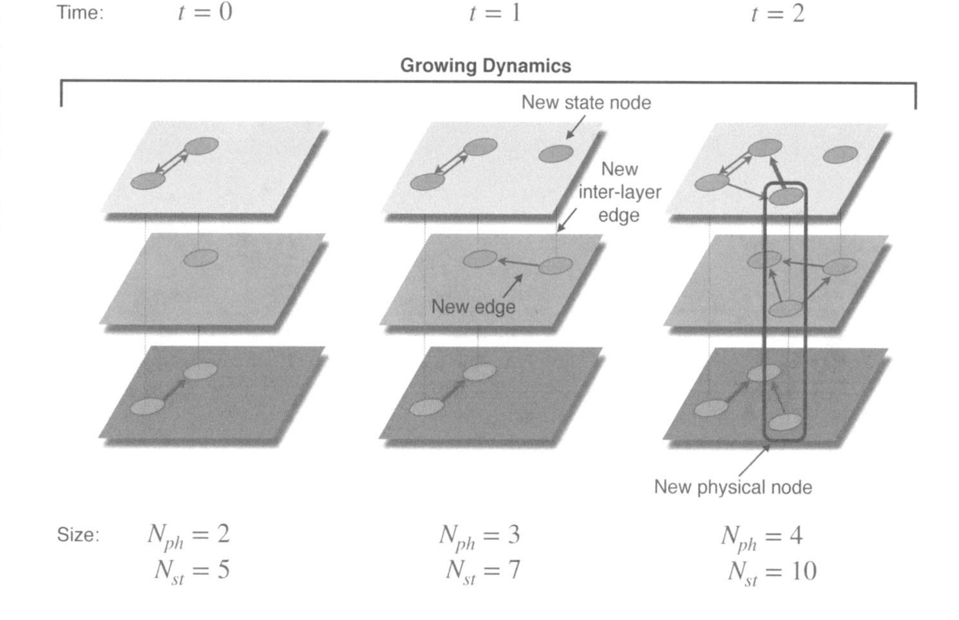

Figure 1.18: Dynamics of growing multilayer networks. The number of physical nodes (indicated by N_{ph}) and state nodes (indicated by N_{st}), as well as the number of edges, increases over time to describe the growth of the system. Figure from [4] under Creative Commons Attribution-ShareAlike 4.0 International License.

in which the critical properties are independent of the other. The two regimes are separated by a *metacritical* point, where a crossover occurs.

In fact, there is a rich spectrum of other dynamical processes that, instead of running on the top of the multilayer structure as the single and coupled dynamics discussed above, define system's change under a specific action. In this framework we can identify i) growth and ii) shrinking dynamics, where the number of nodes, edges and/or layers usually steadily increases or decreases over time, respectively. For sake of completeness, it is worth mentioning another class of dynamics, where growth and shrinking might happen together, causing nodes, edges and layer to be created or to disappear, as in many real systems which adapt to a changing environment. A schematic illustration of such dynamics is shown in Fig. ***1.17*** for a classical network.

The growth dynamics [149] (Fig. ***1.18***) is of special interest and has been investigated to better understand how the system's evolution might induce strong degree correlations across layers which alter its response to spreading dynamics and susceptibility to cascade processes [214], identify condensation phenomena [215] and shed light on the trade-off between efficiency and competition in optimization processes [216].

The study of shrinking dynamics (Fig. ***1.19***) – better known as percolation dynamics in the community of physicists – is of paramount importance to better understand the topological and dynamical robustness of a system, as well as its resilience, to topological perturbations such as targeted attacks to nodes, edges and layers, or their failure due to unpredictable circumstances (e.g., random power outages).

This class of dynamics has attracted the interest of physicists and engineers almost one decade ago, with the pioneering work concerning the study of the consequence of failures, occurring on the electrical grid, on a telecommunication network [217]. The modeling of coupled systems, such as interdependent infrastructures, in a more general framework has been introduced in 2010 [139], putting in evidence that mutual dependence introduces a new level of complexity resulting in novel emergent phenomena, such as the fact that this type of systems are more vulnerable to cascade failures than non-interdependent ones [140], because perturbations – such as random failure of a finite fraction of nodes – on one network can propagate the damage, while being amplified, to nodes in networks that are interdependent, triggering cascade failures which are able to affect the entire system. This iterative process results in a percolation phase transition which is able to isolate the interdependent networks. This

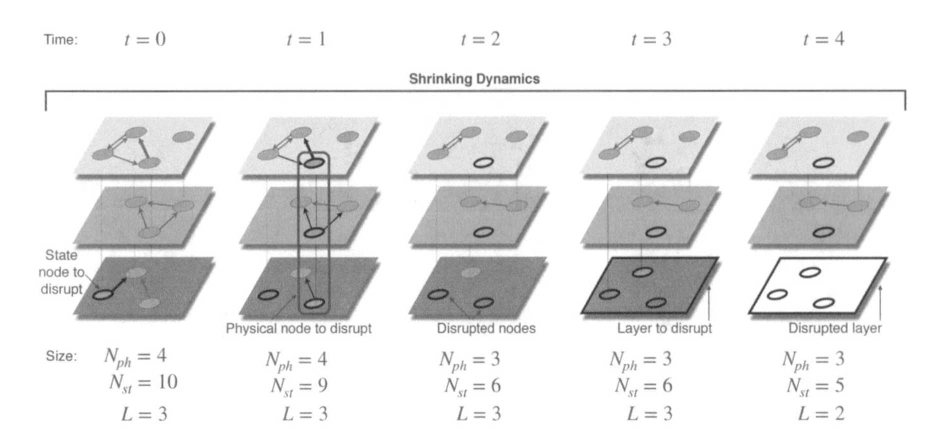

Figure 1.19: Dynamics of shrinking multilayer networks. The number of physical nodes (indicated by N_{ph}) and state nodes (indicated by N_{st}), as well as the number of edges and layers (L), decreases over time to describe the disruption of units and connectivity in the system. Figure from [4] under Creative Commons Attribution-ShareAlike 4.0 International License.

breakthrough opened the doors for theoretical and applied research, investigating under which conditions coupled systems disintegrate their structure or break down their function [141]. It has been shown that by reducing the coupling between networks, the percolation phase transition at the critical point changes from first to second order, with a scaling law characterized by a specific critical exponent [218]. Therefore, the usual percolation theory can be obtained as a limiting case of a more general framework identifying under which conditions cascading failures might be observed and the transition becomes a first-order percolation transition [219]. When the dynamics is studied in the space defined by the fraction of failed nodes and the amount of correlation between high-degree nodes, a triple point is observed, similar to the one separating a nonfunctional phase from two functional phases in liquids [220]. It has been shown that when the correlation between intra- and inter-layer degree is below a certain critical value, the system is driven towards a supercritical regime where dynamical and topological phases are not longer distinguishable [221]. For a class of interdependent networks, hybrid phase transitions depend on directionality within each sub-system, and the overall robustness increases with the in-degree and out-degree correlations [222]. In fact, it is possible to reinforce only a small fraction of nodes to prevent abrupt catastrophic collapses. [223]. Research in this direction is still open and new models are proposed, to account for more realistic conditions, such as the redistribution of flows due to cascade failures [224].

From a mathematical perspective, an interdependent system can be either modeled by a single-layer network with modular structure – where nodes corresponding to different modules are of different type [225] – or by a specific class of multilayer networks, such as an interconnected multiplex, if i) the sub-systems have the same size and ii) there is a one-to-one interdependency between state nodes. In the latter case, the identity of physical nodes is not taken into account and replicas play the role of different units in different modules. This mathematical similarity allowed to study multiplex networks as a special type of interdependent systems and, in fact, the two terms have been used interchangeably in many studies, providing further evidence for the fragility of these systems – facilitated by multiplexity [226, 227] – to avalanche collapse, the emergence of a mutually connected component not depending on the topology of the network of networks [228] and the existence of multiple percolation transitions [229–231]. However, when the physical identity of nodes is taken into account, the concept of overlapping edges (i.e., connections between pairs of nodes which exist simultaneously in multiple layers) [150] is taken into account, it is possible to show that multiplex systems are less fragile than originally thought. The robustness here is boosted either by inter-layer degree correlations [232] or by the existence of the redundant connectivity [225], changing the critical behavior of the percolation in a complex way [233], as well as its structural and dynamical robustness under random failures [18], specially in empirical systems [18, 153, 230].

Capitalizing on these results, it is natural to wonder if, and under which conditions, it is possible to find a minimal set of nodes that, once removed, dismantle the system into fragmented, non-extensive, disconnected clusters, unable to allow information to be exchanged among system's units. Identification of this set is of crucial importance for a variety of applications, from the determination of more effective and less expensive immunization strategies to fight the spreading of epidemic diseases to the maximization of information diffusion from a few spreaders, known as *influencers* in (computational) social science. This question, also known as *optimal percolation*, is almost two decades old, from the pioneering work on error and attack tolerance of complex networks [103] (see [54] for a review) to the more recent dismantling techniques [121, 123], but only recently the challenge has been tackled for multilayer systems. In fact, it has been shown that classical approaches such as damaging 2–cores, i.e. decycling, are less effective in destroying the giant connected component than using strategies which are inherently multilayer [234], e.g., using the multiplex degree [235].

These studies are relevant also for designing optimal recovery strategies, such as damage repair [236], or for enhancing the robustness of complex systems [237]. At variance with interdependent networks, a common feature of multiplex systems seems to be a higher resilience than their individual layer as quantified, for instance, by a larger navigability in transportation and communication networks [18]. Similar techniques are also used to better understand other real multilayer systems. An interesting example is given by the interdependency between human activity and environment in social-ecological networks, altering both the climate and the response of humankind to its dramatic changes. By considering a variety of plausible scenarios, including the ones related to global warming, it is possible to quantify the robustness of certain communities – such as the ones in northern Alaska villages, characterized by mixed subsistence-cash economy – to the corresponding resource depletion and to social changes, discovering that the latter play a more fundamental role for the connectedness of these systems [8].

1.5 muxViz : A tool for data science of multilayer networks

The area of multilayer networks has been rapidly growing in less than a decade when it attracted the interest of researchers and practitioners in hard science. Nowadays, the available number of algorithms for modeling and analyzing this type of system is large.

However, these new analytical tools are often difficult to be used in practice because of at least one of the following issues:

- The algorithm is difficult to implement and source code is not available;
- Source code is available in a very specific programming language, not a mainstream one (e.g., R or Python);
- Source code is in a mainstream programming language, but it relies on specific packages or libraries which are difficult to install or mostly undocumented;
- Source code of different tools are available either in different programming languages or are based on different packages or libraries.

The most common problem has been the lack of a unified computing framework, where multilayer networks could be represented, manipulated, analyzed and visualized without too much efforts from the researcher or the practitioner on porting existing code or merging different codes. muxViz [19] is a free and

Figure 1.20: Data page in `muxViz` as rendered by any standard Web browser. The visualization is interactive and information about each single data set can be easily retrieved and reported.

open source project, publicly available (https://github.com/manlius/muxViz) under GNU Public License v3 (https://www.gnu.org/licenses/gpl-3.0.en.html), conceived and developed in 2013 to fill this gap.

Nowadays, **`muxViz`** counts on a community of more than 600 researchers and practitioners, and it is considered as the most complete platform for analyzing and visualizing multilayer networks. The framework is based on a library developed in R language[7] and is powered by a Graphical User Interface (GUI) exploiting the Shiny app technology[8]. The GUI is user-friendly (Fig. ***A.2***) and runs in any Web browser to provide access to many customizable graphic and analytical options to analyze and visualize complex multilayer networks.

The great advantage of **`muxViz`** is that it is designed for being used either by users with coding experience (specifically, R language) able to build complex data analytics from available functions or by users with no coding skills through the interface. From **`muxViz`** it is possible to explore an online repository of multilayer networks (Fig. ***1.20***) with dozens of publicly available fully referenced data sets (Fig. ***1.21***). However, a dedicated utility to import downloaded files and to convert them to **`muxViz`** format is not yet available.

In the next chapter we will specify the details about the software environment required to install **`muxViz`** and we will describe how to format multilayer network data to be imported into the framework.

[7] https://www.r-project.org/
[8] https://shiny.rstudio.com/

Figure 1.21: The Data page includes a form to navigate through available data sets and search for specific ones.

Show 25 entries Search:

Name	Type	Directed	Weighted	Layers	Nodes	Edges	Reference
HIGGS TWITTER	social	yes	yes	4	456631	16070185	M. De Domenico, A. Lima, P. Mougel and M. Musolesi. The Anatomy of a Scientific Rumor. (Nature Open Access) Scientific Reports 3, 2980 (2013)
LONDON TRANSPORT	transportation	no	yes	13	369	503	Manlio De Domenico, Albert Solé-Ribalta, Sergio Gómez, and Alex Arenas. Navigability of interconnected networks under random failures. PNAS 111, 8351-8356 (2014)
EU-AIR TRANSPORT	transportation	no	no	37	450	3588	Alessio Cardillo, Jesús Gómez-Gardenes, Massimiliano Zanin, Miguel Romance, David Papo, Francisco del Pozo and Stefano Boccaletti. Emergence of network features from multiplexity. Scientific Reports 3, 1344 (2013)
CS-AARHUS	social	no	no	5	61	620	Matteo Magnani, Barbora Micenkova, Luca Rossi. Combinatorial Analysis of Multiple Networks. arXiv:1303.4986 (2013)
CKM PHYSICIANS INNOVATION	social	yes	no	3	246	1551	J. Coleman, E. Katz, and H. Menzel. The Diffusion of an Innovation Among Physicians. Sociometry (1957) 20:253-270
KAPFERER TAILOR SHOP	social	yes	no	4	39	1018	Kapferer B. (1972). Strategy and transaction in an African factory
KRACKHARDT HIGH TECH	social	yes	no	3	21	312	D. Krackhardt. Cognitive social structures. Social Networks (1987), 9, 104-134

Chapter 2 | Multilayer Networks: Overview

Contents

In the previous chapter, we have provided a very quick introduction to network models of complex systems, to briefly motivate the needing for more general models – i.e., multilayer networks – to be used to better understand the structure and dynamics of empirical systems and systems of systems.

In this chapter we will focus our attention on the theoretical background of multilayer analysis and visualization within the framework of muxViz. Therefore, it is worth remarking that this chapter does not intend to provide an exhaustive survey of available analytical techniques: instead, it is devoted to introduce the reader to the several analytical techniques that are available either through the graphical user interface (GUI) of muxViz or through the scripting functions which build the accompanying multilayer network library (LIB).

2.1 Multilayer network models

muxViz allows to import all multilayer network models introduced in the previous chapter and summarized in Fig. *1.11*. The graphical user interface (Fig. *2.1*) is designed to follow a linear workflow:

1. Select the multilayer model to use;
2. Specify the basic properties of the model (is it weighted? Is it directed?);
3. Specify the type of inter-layer connectivity to apply automatically in the case of edge-colored input[9]:
   - **Ordinal**: only adjacent layers are interconnected, like in an undirected chain;
   - **Categorical**: all layers are interconnected with each other, like in an undirected clique;
   - **Temporal**[10] **(acyclic)**: like ordinal, but only in one direction;
   - **Temporal**[11] **(cyclic)**: like ordinal, but only in one direction, with the last layer interconnected to the first one.
4. Select the input files of the data and import them.

[9] This option is required because the tensorial framework can be safely used only to analyze systems with interconnected or interdependent layers.

[10] In muxViz it is available only through the standalone library.

[11] In muxViz it is available only through the standalone library.

M. De Domenico, *Multilayer Networks: Analysis and Visualization*, https://doi.org/10.1007/978-3-030-75718-2_2

2.2 Representing multilayer networks

In general, the mathematical representation of multilayer networks is given by rank–4 tensors. In the previous chapter we have introduced the *multilayer adjacency tensor* $M^{i\alpha}_{j\beta}$ which can be thought as an hypermatrix with four indices. The tensorial representation of multilayer networks allow to build a powerful mathematical framework for their analysis, as we will see in the next sections of this chapter. In practice, this object encodes the interaction strength between a node i in layer α and a node j in layer β: note that relationships can be also directed and weighted, without requiring to change the mathematical framework [129].

The framework is still valid even if all nodes do not exist in all layers simultaneously: this might correspond to the case of a bus stop present where a nearby tube station is missing in a urban multimodal transportation network, or to an individual with an account in some social networks (e.g., Twitter) but not in others (e.g., Facebook). From a practical point of view, the absence of a node in a specific layer can be encoded by padding that layer with an *empty node*, i.e., a node with no edges. The absence of edges is encoded by assigning the value 0 to the corresponding entries in the multilayer adjacency tensor and, if this is the case, one should be careful to normalize adequately the network descriptors to account for this padding procedure [129].

Section A.5 provides details about different ways of storing information contained in a multilayer adjacency tensor: depending on the complexity of the network model it is possible to use more efficient ways to store multilayer interactions. For instance, a non-interconnected multiplex/edge-colored network does not require to store inter-layer connectivity information: this reduction of complexity can be easily exploited to encode relationships using only three indices, one to identify the layer and two to identify the corresponding pair of nodes. In fact, this type of networks can be easily stored in edge lists with just 4 columns: one for the layer, two for the nodes and one for the weight of the connection. This can be further reduced to 3 columns in case of unweighted networks. Non-interconnected multiplex networks can be also conveniently stored in multiple edge lists, each one corresponding to a specific layer and consisting

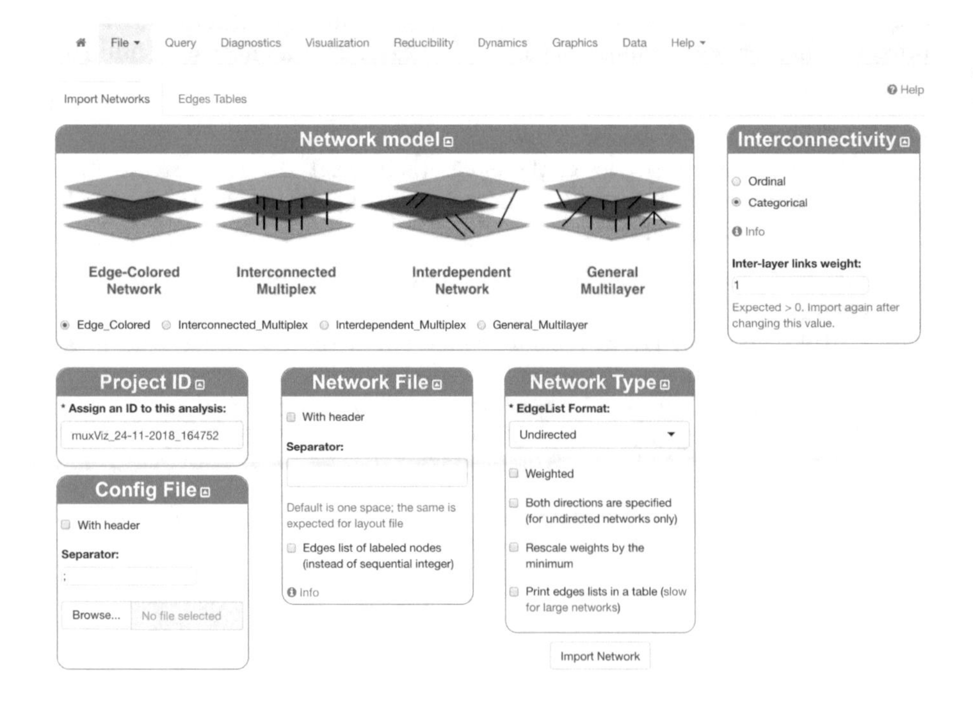

Figure 2.1: Import page in `muxViz` as rendered by any standard Web browser.

of 2 or 3 columns, in the case of unweighted and weighted networks, respectively.

▸ Code snippet 2.1
example_plot_edgecolored.R

More generally, multilayer networks are uniquely represented by edge lists with 5 columns: one for the origin layer and one for the origin node, one for the destination layer and the destination node, and one for the weight of the connection.

There are some ways to generate plausible multilayer networks, more or less sophisticated [238, 239], which account or not for correlations across layers. In `muxViz` one can develop her own generative models or use existing generative algorithms which falls into the class of methods where each layer is generated independently from the others.

In Fig. *2.2* we show the result of a simple algorithm to generate three highly correlated layers with a Barabasi-Albert topology. Figure *2.3* shows a similar example, where the organization into multilayer communities is used instead.

▸ Code snippet 2.2
example_plot_edgecolored_heatmap_6panels.R

2.3 Fundamental tensors

We have already encountered the first fundamental tensor, namely the ***multilayer adjacency tensor*** $M^{i\alpha}_{j\beta}$, in the previous chapter. This object is a genuine tensor (see Box 2.3.1) which can be decomposed into four main tensors (see Box 2.3.2), accounting for:

1. **Intra-layer interactions**: which can be further distinguished into
   a. **self-interactions**: from a node to itself;
   b. **endogeneous interactions**: between different nodes within the same layer;
2. **Inter-layer interactions**: which can be further distinguished into
   a. **intertwining**: from a node to its replicas in other layers;
   b. **exogenous interactions**: between different nodes across different layers.

Box 2.3.1: The adjacency tensor of a multilayer network

In Box 1.1.2 we have shown the tensorial nature of adjacency tensors representing monoplex networks. In the same spirit, we introduce the vectors $e^{\alpha}(p)$ $(\alpha, p = 1, \dots, L)$ of the canonical basis in the space $\mathbb{R}^L$, being L the number of layers. Note that, for sake of simplicity, here we want to make a notational difference between components related to nodes and layers: we use Latin letters for indices related to nodes and Greek letters for indices related to layers. Ad-

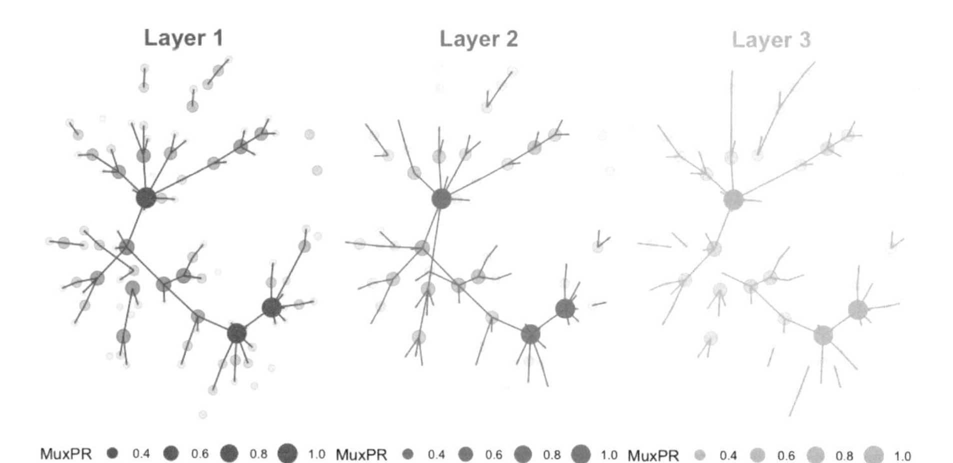

Figure 2.2: Three layers with a Barabasi-Albert structure, consisting of 100 nodes not necessarily connected in all layers. Colors encode distinct layers, while node size encode its PageRank versatility. This type of visualization is available only for the LIB version.

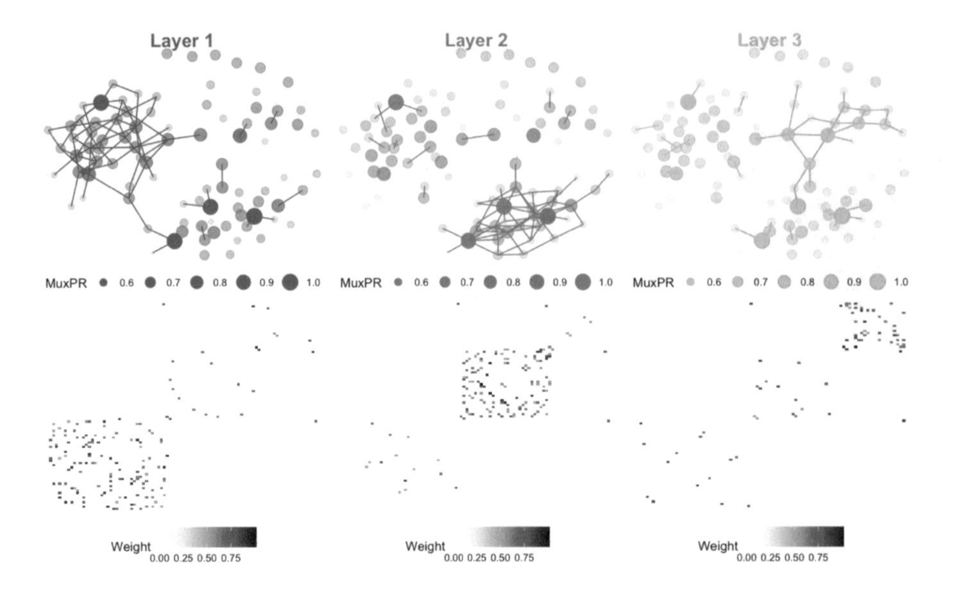

Figure 2.3: as in Fig. ***2.2***, but for a system with an organization in multilayer groups (top panels). Additionally, here we show the heatmap corresponding to each layer, to highlight the underlying block structure. This type of visualization is available only for the LIB version.

ditionally, to indicate the p–th element of a set, like the p–th canonical vector, we use Latin letters.
It follows that the 2nd-order tensors $E^{\alpha}_{\beta}(pq) = e^{\alpha}(p)e_{\beta}(q)$ represent the canonical basis of the space $\mathbb{R}^{L\times L}$. Similarly to the case of monoplex networks, it can be shown [129] that

$$M^{i\alpha}_{j\beta} = \sum_{a,b=1}^{N}\sum_{p,q=1}^{L} w_{ab}(pq)e^{i}(a)e_{j}(b)e^{\alpha}(p)e_{\beta}(q) \quad (2.1)$$

defines a multilayer object in terms of the Kronecker product of canonical vectors in a higher dimensional space. As in Box 1.1.2, we can show that this object transforms like a tensor under a change of coordinates:

$$\begin{aligned} M'^{i\alpha}_{j\beta} &= \sum_{a,b=1}^{N}\sum_{p,q=1}^{L} w_{ab}(pq)Q^{i}_{k}e^{k}(a)(Q^{-1})^{l}_{j}e_{l}(b)\tilde{Q}^{\alpha}_{\gamma}e^{\gamma}(p)(\tilde{Q}^{-1})^{\delta}_{\beta}e_{\delta}(q) \\ &= Q^{i}_{k}\tilde{Q}^{\alpha}_{\gamma}M^{k\gamma}_{l\delta}(Q^{-1})^{l}_{j}(\tilde{Q}^{-1})^{\delta}_{\beta}. \end{aligned} \quad (2.2)$$

Box 2.3.2: Structural SNXI decomposition of the multilayer adjacency tensor

It is possible to isolate four different tensors from $M^{i\alpha}_{j\beta}$, each one corresponding to a specific set of structural relationships. To avoid any confusion with the tensorial object, in the following we indicate its components by $m^{j\beta}_{i\alpha}$ ($i,j = 1,2,\ldots,N$ and $\alpha,\beta = 1,2,\ldots,L$). We use δ^{j}_{i} and δ^{β}_{α}, respectively, to indicate the Kronecker delta function for indices corresponding to nodes and layers. The four different contributions to node–node relationships within and across layers of the multilayer network can be summarized as

$$\begin{aligned} m^{j\beta}_{i\alpha} &= \underbrace{m^{j\beta}_{i\alpha}\delta^{\beta}_{\alpha}\delta^{j}_{i} + m^{j\beta}_{i\alpha}\delta^{\beta}_{\alpha}(1-\delta^{j}_{i})}_{\text{intra-layer relationships}} + \underbrace{m^{j\beta}_{i\alpha}(1-\delta^{\beta}_{\alpha})\delta^{j}_{i} + m^{j\beta}_{i\alpha}(1-\delta^{\beta}_{\alpha})(1-\delta^{j}_{i})}_{\text{inter-layer relationships}} \\ &= \underbrace{m^{i\alpha}_{i\alpha}}_{\text{self-relationships}} + \underbrace{m^{j\alpha}_{i\alpha}}_{\text{endogenous}} + \underbrace{m^{j\beta}_{i\alpha}}_{\text{exogenous}} + \underbrace{m^{i\beta}_{i\alpha}}_{\text{intertwining}} \\ &= \mathbb{S}_{i\alpha}(M) + \mathbb{N}^{j}_{i\alpha}(M) + \mathbb{X}^{j\beta}_{i\alpha}(M) + \mathbb{I}^{\beta}_{i\alpha}(M). \end{aligned} \quad (2.3)$$

Equation (2.3) defines the "structural $\mathbb{SNXI}$ decomposition" of the multilayer adjacency tensor M. Different types of multilayer networks, as the ones shown in Fig. ***1.11***, arise from contributions of different $\mathbb{SNXI}$ components in the tensorial representation of a multilayer network.

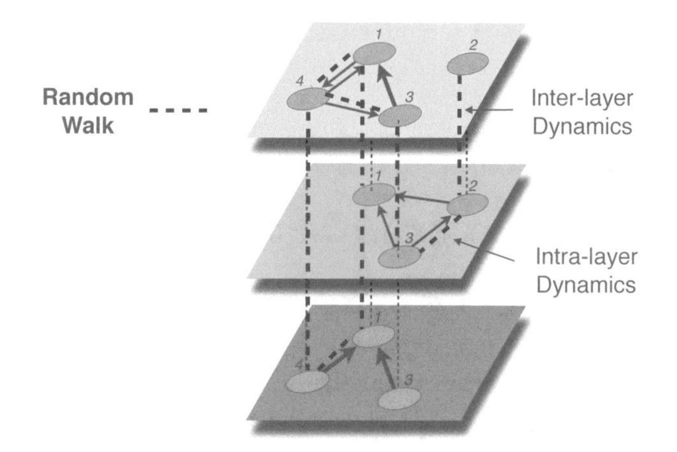

Figure 2.4: Illustration of a random walk typical multilayer network. Random walks are used for a wide variety of applications, from modeling random searches, exploration and navigability of networked systems to modeling information diffusion. Once a random walker, i.e. an agent, is placed in one node of a specific layer, it can perform 4 different actions: stay in the same node, jump to a neighbor in the same layer, switch to another state node or jump anywhere else, according to its transition rules, encoded in the corresponding transition tensor (see the text for further details). The collection of nodes and edges visited sequentially defines the walk, here represented by a dashed line. Figure from [4] under Creative Commons Attribution-ShareAlike 4.0 International License.

It is constructive to introduce some tensors with different rank, widely used in the reminder of this chapter. For instance, we will use $e^i(a)$ to indicate the rank–1 canonical tensors in the space of nodes, corresponding to vectors with dimension N with all entries equal to 0 except for the a–th one, equal to 1. Similarly, we define $e^\alpha(p)$ as the rank–1 canonical tensors in the space of layers, corresponding to vectors with dimension L. Higher order canonical tensors are obtained by the product of lower order ones. For instance, we will use $E^i_j(ab) = e^i(a)e_j(b)$ to indicate rank–2 canonical tensors in the space of nodes, encoding $N \times N$ matrices with all entries equal to 0 except for the one corresponding to the a–th row and the b–th column, which is equal to 1. Similarly, we will use $E^\alpha_\beta(pq) = e^\alpha(p)e_\beta(q)$ to indicate rank–2 canonical tensors in the space of layers, corresponding to $L \times L$ matrices. More generally, the multilayer canonical tensor is a rank–4 object indicated by $E^{i\alpha}_{j\beta}(ab;pq) = E^i_j(ab)E^\alpha_\beta(pq)$.

Another useful battery of objects is given by the 1–tensors, which are tensors with all components equal to 1. This is the case of u^i and $U^i_j = u^i u_j$ – rank–1 and rank–2, respectively – in the space of nodes, and, similarly, u^α and $U^\alpha_\beta = u^\alpha u_\beta$ in the space of layers. The multilayer 1–tensor is $U^{i\alpha}_{j\beta} = U^i_j U^\alpha_\beta$.

The Kronecker tensor is another widely used object. In the space of nodes it is indicated by δ^i_j and its components are equal to 1 if $i = j$ and equal to 0 otherwise. Similarly, in the space of layers we make use of δ^α_β. In the multilayer space, this object is indicated by $\delta^{i\alpha}_{j\beta}$. An important tensor, defined in terms of the previous ones, is $F^{i\alpha}_{j\beta} = U^{i\alpha}_{j\beta} - \delta^{i\alpha}_{j\beta}$, representing a complete multilayer network without self-edges: this plays an important role in quantifying triadic closure in multilayer systems.

2.4 Dynamical processes

Single dynamics (see Fig. ***1.16***) can be used to define a broad class of multilayer network descriptors, from centrality measures to system's organization in functional modules and navigability. Here, we briefly introduce one of the simplest dynamics used for this purpose: diffusion.

Diffusive processes have been successfully used in a variety of applications, from information spreading in socio-technical networks to the spreading of infectious diseases in social systems and the synchronization dynamics of oscillators (see [145] for a recent review). In the following, we will refer to *information diffusion* to indicate the spreading dynamics of an agent, being regardless if it is a pathogen or a meme.

Multilayer networks allow information to diffuse within or across layers, by means of intra- and inter-layer connectivity. Let $X_{i\alpha}(t)$ indicate the state tensor of replica nodes – encoding the state of the information at node i in layer α – at time t. In our framework, this state tensor can be imagined either as a rectangular matrix with dimension $N \times L$ or as a (supra-)vector with dimension $1 \times NL$. The equation governing the *continuous* dynamics of changes in this state tensor is given by

$$\frac{dX_{j\beta}(t)}{dt} = M^{i\alpha}_{j\beta} X_{i\alpha}(t) - M^{i\alpha}_{k\gamma} U_{i\alpha} E^{k\gamma}(i\beta) X_{i\beta}(t)$$
$$= -L^{i\alpha}_{j\beta} X_{i\alpha}(t), \qquad (2.4)$$

where $U_{i\alpha} = u_i u_\alpha$, $E^{k\gamma}(i\beta) = e^k(i) e^\gamma(\beta)$ and $L^{i\alpha}_{j\beta}$ is the multilayer Laplacian tensor. This type of dynamics is suitable to model a continuous diffusion of information from a node to its neighbors: for instance, like water flowing between neighbors (nodes) through pipes (edges). It has been extensively shown that the mathematical properties of the solution of the diffusion equation, given by $X_{j\beta}(t) = X_{i\alpha}(0) e^{-L^{i\alpha}_{j\beta} t}$, can be characterized from the analysis of the eigenvalue spectrum of the Laplacian tensor, but we refer the interested read to the relevant literature for further details [129, 133, 158, 161].

More often, the diffusion dynamics of interest for application is *discrete* in time. This is the case when information, in a single time step, can be transmitted from a node to one of its replicas or its neighbors only. A canonical example of such a process is a random walk [240, 241], corresponding to a Markovian process on the network. Random walks are among the processes most used to approximate more complex dynamics because of their analytical tractability. For a thorough discussion about this topic, we refer to the recent review by Masuda, Lambiotte and Porter [242].

In classical networks, the simplest rule for a random walk is local: a walker can only *jump* from a node to another one in its close neighborhood. In multilayer networks, the same walker has multiple options: on the one hand it can *jump* between nodes within the same layer through intra-layer edges, while on the other hand it can *switch* between state nodes through inter-layer edges (see Fig. *2.4*). The combination of jumps and switches allows the walker to explore the whole multilayer structure. The rules can be even more complicated, allowing to define a broad set of network descriptors as we will see later in this chapter. It is worth remarking here that such rules can be elegantly encoded into a specific tensor, named *multilayer transition tensor*, which governs the evolution of walk by means of a master equation [18, 129].

Let us indicate the multilayer transition tensor by $\mathcal{P}^{i\alpha}_{j\beta}$: this encodes the probability that a walker located in node i in layer α can jump or switch to node j in layer β $(i, j = 1, 2, ..., N$. Let $p_{i\alpha}(t)$ denote the probability of finding the random walker in node i of layer α at time t: the master equation, in compact form and adopting Einstein summation convention, reads

$$p_{j\beta}(t + \Delta t) = \mathcal{P}^{i\alpha}_{j\beta} p_{i\alpha}(t), \qquad (2.5)$$

and gives the probability to find the random walker in node j of layer β at the next time step $t + \Delta t$. Note that, usually, $\Delta t = 1$. It can be useful to expand this equation to highlight the contribution of jumps and switches in the dynamics:

$$p_{j\beta}(t+\Delta t) = \underbrace{\mathcal{P}^{j\beta}_{j\beta} p_{j\beta}(t)}_{\text{stay}} + \underbrace{\sum_{\substack{\alpha=1 \\ \alpha\neq\beta}}^{L} \mathcal{P}^{j\alpha}_{j\beta} p_{j\alpha}(t)}_{\text{switch}} + \underbrace{\sum_{\substack{i=1 \\ i\neq j}}^{N} \mathcal{P}^{i\beta}_{j\beta} p_{i\beta}(t)}_{\text{jump}} + \underbrace{\sum_{\substack{\alpha=1 \\ \alpha\neq\beta}}^{L} \sum_{\substack{i=1 \\ i\neq j}}^{N} \mathcal{P}^{i\alpha}_{j\beta} p_{i\alpha}(t)}_{\text{switch and jump}} .$$

The continuous-time approximation of the master equation allows one to write it as the differential equation

$$\frac{dp_{j\beta}(t)}{dt} = -\tilde{L}^{i\alpha}_{j\beta} p_{i\alpha}(t), \tag{2.6}$$

where $\tilde{L}^{i\alpha}_{j\beta} = \delta^{i\alpha}_{j\beta} - \mathcal{P}^{i\alpha}_{j\beta}$ is known as the normalized multilayer Laplacian tensor. The solution of this equation is formally equivalent to the one of diffusion equation, the difference being in the the Laplacian tensor defining the propagator of the dynamics.

More complex dynamics on the top of multilayer structures can be defined as well, but they are well beyond the scope of this book. We refer to [56, 57, 142, 143, 145, 243] for reviews and more technical details about more complex dynamics, as well as to Box 2.4.1.

Box 2.4.1: Dynamical SNXI decomposition of the multilayer adjacency tensor

We have seen in Box 2.3.2 that the structure of the multilayer adjacency tensor can be decomposed to highlight the contribution of four different tensors. Similarly, a broad variety of multilayer dynamical processes can be understood in terms of a dynamical $\mathbb{SNXI}$ decomposition, to describe dynamics on multilayer networks. The approach is similar in spirit to the one used by Golubitsky, Stewart and Török to model coupled cell networks [244], where the difference is that here the structural and dynamical effects are explicitly separated. Let $x^{[\ell]}_{i\alpha}$ (where $\ell \in \{1, 2, \dots, C\}$) denote the ℓ-th component of a C-dimensional vector $x_{i\alpha}$ that represents the state of node i in layer α. Indicating by $X(t) \equiv (x_{11}, x_{21}, \dots, x_{N1}, x_{12}, x_{22}, \dots, x_{N2}, \dots, x_{1L}, x_{2L}, \dots, x_{NL})$, the most general (and possibly nonlinear) dynamics governing the evolution of each state is given by the systems of equations

$$\begin{aligned}
\dot{x}_{i\alpha}(t) = F_{i\alpha}(X(t)) &= \sum_{\beta=1}^{L}\sum_{j=1}^{N} f^{j\beta}_{i\alpha}(X(t)) \\
&= \underbrace{\sum_{\beta=1}^{L}\sum_{j=1}^{N} f^{j\beta}_{i\alpha}(X(t))\,\delta^{\beta}_{\alpha}\delta^{j}_{i} + \sum_{\beta=1}^{L}\sum_{j=1}^{N} f^{j\beta}_{i\alpha}(X(t))\,\delta^{\beta}_{\alpha}(1-\delta^{j}_{i})}_{\text{intra-layer dynamics}} \\
&\quad \underbrace{+ \sum_{\beta=1}^{L}\sum_{j=1}^{N} f^{j\beta}_{i\alpha}(X(t))(1-\delta^{\beta}_{\alpha})\,\delta^{j}_{i} + \sum_{\beta=1}^{L}\sum_{j=1}^{N} f^{j\beta}_{i\alpha}(X(t))(1-\delta^{\beta}_{\alpha})(1-\delta^{j}_{i})}_{\text{inter-layer dynamics}} \\
&= \underbrace{f^{i\alpha}_{i\alpha}(X(t))}_{\text{self-interaction}} + \underbrace{\sum_{j\neq i} f^{j\alpha}_{i\alpha}(X(t))}_{\text{endogenous interaction}} + \underbrace{\sum_{\beta\neq\alpha}\sum_{j\neq i} f^{j\beta}_{i\alpha}(X(t))}_{\text{exogenous interaction}} + \underbrace{\sum_{\beta\neq\alpha} f^{i\beta}_{i\alpha}(X(t))}_{\text{intertwining}} \\
&= \mathbb{S}_{i\alpha}(X(t)) + \mathbb{N}_{i\alpha}(X(t)) + \mathbb{X}_{i\alpha}(X(t)) + \mathbb{I}_{i\alpha}(X(t)),
\end{aligned} \tag{2.7}$$

where the different contributions of intra-layer and inter-layer dynamics have been decoupled. This decomposition allows to classify different dynamical processes in terms of the corresponding dynamical $\mathbb{SNXI}$ components.

Chapter 3 Multilayer Analysis: Fundamentals and Micro-scale

Contents

ONCE the connectivity of nodes and layers is appropriately encoded in the multilayer adjacency tensor, it is possible to define a new set of descriptors useful to characterize the multilayer structure of a system. However, it is worth stressing the fact that naive generalizations of existing measures can lead to quantitatively different results, often incorrect or misleading [143]. Here, we are going to describe the measures included in `muxViz`.

3.1 Descriptive statistics per layer

The rank–2 adjacency tensors corresponding to layers can be obtained by projecting the multilayer adjacency over adequate canonical rank–2 tensors. Let us start with an example on problems with lower dimension.

Let v_i a vector whose components provide some type of information about nodes in a classical network: this vector has dimension N and we are interested in extracting the single component, i.e. a scalar, which correspond to the a–th node. This operation of extraction is obtained by projecting the vector onto another vector: the canonical one corresponding to the a–th node, which is $e^i(a)$. The projection is an internal product and, in fact, it corresponds to $v_i e^i(a)$.

Similarly, we can act on higher-order tensors, such as the rank–2 adjacency matrix A^i_j. Let us assume that we know the b–th column, which is a vector encoding the incoming connections from the rest of the network to node b. Once again, we project on the corresponding canonical vector: i.e., the result is the vector $A^i_j e^j(b)$.

The same approach holds for even higher dimensions, such in the case of the multilayer adjacency tensor. If we are interested in extracting the layer p, which is now a rank–2 tensor, we need to project on the corresponding rank–2 canonical tensor: $M^{i\alpha}_{j\beta} E^{\beta}_{\alpha}(pp) = G^i_j(p)$.

Once we obtain the rank–2 adjacency tensor of each layer, we can perform some standard analysis of the corresponding classical networks, keeping in mind that no multilayer effects can be discovered by using this approach. A thorough analysis of single-layer descriptors is beyond the scope of `muxViz` and this book (we refer the reader to the excellent books on network science briefly mentioned in the introductory chapter), therefore we limit here to mention that the GUI allows one to analyze the density distribution of:

- **Nodes**: the number of non-isolated nodes per layer;

M. De Domenico, *Multilayer Networks: Analysis and Visualization*, https://doi.org/10.1007/978-3-030-75718-2_3

- **Edges**: the number of edges per layer;
- **Density**: the ratio between the number of edges and nodes to estimate the average degree per layer[12];
- **Components**: the number of connected components per layer;
- **Diameter**: the size of the diameter – i.e., the longest shortest path – per layer;
- **Mean Path Length**: the average path length – i.e., the sample mean of all shortest paths – per layer.

[12] Note that it is common to define the density by the ratio between the number of edges and the maximum possible number of edges, i.e. $N(N-1)/2$ for undirected networks without self-loops, $N(N-1)$ for directed networks without self-loops and N^2 for directed networks where self-loops are allowed.

Once the adjacency tensor of each layer is available, one can perform any other type of analysis through the R environment and the LIB. Depending on the data format, it can be more efficient to use an approach instead of another one if the interest is in the single-layer analysis. For instance, if each layer is stored in a separate file, then it is more efficient to import each layer separately and perform the network analysis without passing through the LIB, which instead is designed to work with multilayer network data.

3.2 Aggregate network

The aggregate network is, as per its name, an aggregate representation of a multilayer system [129]. There are multiple ways of aggregating a multilayer network to a single layer, each one depending on the application of interest. There are different reasons for using aggregated networks, such as for instance to: i) reduce the dimensionality of the system to avoid the necessity to work with large tensors; ii) filter out noisy information.

Very often, aggregation has been used to describe with a static network the complex structure of a system changing over time. In fact, special cases of multilayer adjacency tensors are time-dependent (i.e., "temporal") networks [129, 143], although analyzing this type of systems requires to be very careful because of the presence of a privileged direction in the structural representation, due to the arrow of time. Historically, temporal networks have been studied well before the development of the mathematical formulation of multilayer networks and can count on a very active community of researchers who introduced a broad spectrum of network descriptors(see [56, 57] for a review). However, unifying the methodologies developed for the analysis of temporal and static multilayer networks is still a challenging problem whose solution is fundamental to advance network science towards a consistent mathematical theory.

Whatever the reason for aggregating information, it is worth remarking here that analyzing the aggregate network instead of the whole multilayer system can lead to spurious and misleading results, especially when layer-layer interactions are important and correlations can not be neglected. However, this is not always the case and, as we will see later in this chapter, approaches to reduce the dimensionality of multilayer networks have been proposed and, sometimes, they suggest that the aggregate network is representative of the system.

Nevertheless, the comparison between results obtained from multilayer analysis and the analysis of the aggregate is now considered as a standard benchmark in the community, mostly to answer some important questions such as *"is the multilayer representation needed?"* or *"can I gain the same insights from a classical analysis of layers separately or of the aggregate network?"*.

The simplest operations that can be applied to obtain an aggregate network are (Fig. ***3.1***):

- **Sum**: the rank–2 adjacency tensors corresponding to each layer are summed up to build a new rank–2 adjacency tensor with the same dimensions (i.e., $N \times N$) representing a new network where the link between nodes i and j is weighted by the sum of the weights of their links across layers;

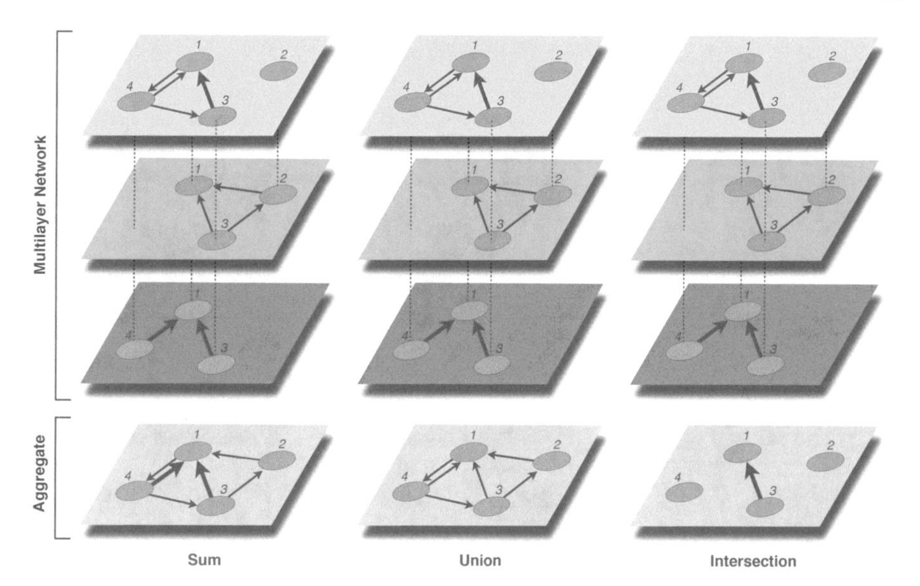

Figure 3.1: Different aggregate representations of the same multilayer network. See the text for further details. Figure from [4] under Creative Commons Attribution-ShareAlike 4.0 International License.

- **Average**: like in the previous case, but the total weight is divided by the number of layers L, providing an average interaction strength instead of a total one;
- **Union**: for each pair of nodes i and j a link exists in the aggregate representation if they are linked in at least one layer. In this case, it is not evident how to deal with weighted networks and the result is often an unweighted network;
- **Intersection**: like in the previous case, but considering the intersection across layers, i.e. the link between nodes i and j exists in the aggregate if and only if they are linked in all layers simultaneously.

For the analysis, `muxViz` in general makes use of the representation aggregated by the sum, because it preserves the overall link weight. For the visualization, instead, the operations corresponding to union and intersection are available and can be useful to build better layouts to visualize the network.

The aggregate network G^i_j is a monoplex and, in the case of sum, it is obtained by contracting the layer indexes of the multilayer adjacency tensor, i.e., $G^i_j = M^{i\alpha}_{j\alpha}$. This type of aggregation washes out the information about inter-layer connectivity, i.e., the weight of inter-layer links are not accounted for. When this information is important for the analysis of interest, it is possible to account for inter-layer connectivity by contracting the multilayer adjacency tensor with the rank–2 1–tensor U^{β}_{α}, to obtain the corresponding aggregate representation: $\bar{G}^i_j = M^{i\alpha}_{j\beta}U^{\beta}_{\alpha}$.

3.3 Layer-layer correlations

The coupling between layers induces dramatic changes in the structural properties of multilayer networks. This coupling can be of different types, depending on the presence or absence of inter-layer links, i.e., if our system is interconnected or not. In the case of non-interconnected multiplex systems, many measures have been introduced to study and quantify correlations among layers (see, for instance, Refs. [151, 152, 245, 246]). However, here we are more interested in structures characterized by inter-layer connectivity, whose presence produces several interesting structural and dynamical phenomena. In social networks, for instance, an inter-layer link allows one to model individual's self-reinforcement in opinion dynamics [213]. Another interesting application is in transportation engineering, where multimodal systems allow to model differ-

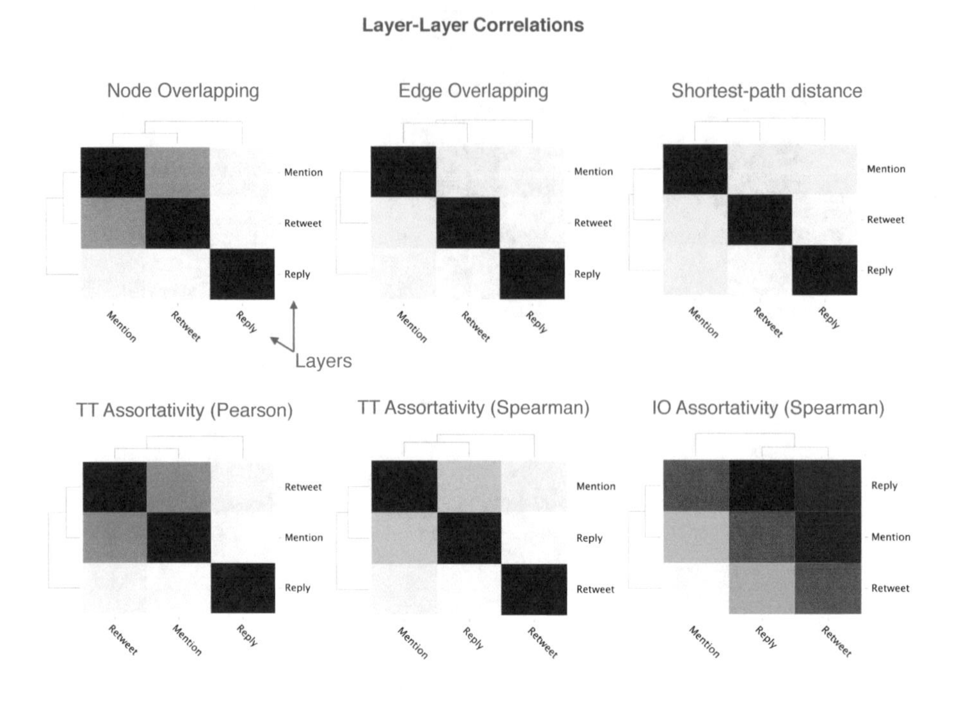

Figure 3.2: Different measures of layer-layer correlations for the analysis of an empirical multilayer social system. The data encodes online interactions between users in Twitter during the Conference on Complex Systems held in Cancun (Mexico) in 2017. The layers encode different social actions in the system: *Reply* (who replies to an existing messages posted by someone else), *Mention* (who mentions someone else, regardless of the existence of a previous message) and *Retweet* (who endorses the message posted by someone else). The three actions have different social meaning and it is interesting to study their correlation. Here, all the layer-layer correlation measures available in **muxViz** are used (see the text for details), with colors encoding the strength of the correlation (the darker the higher). Correlation values have been used to cluster together the layers, to gain some insight about the hierarchy of the social actions they encode.

ent transportation modes serving the same geographical areas (e.g., within a city or a whole region) and the weight of inter-layer connections can be tuned encode the cost (e.g., economic, temporal, *etc*) to switching between two modes [18, 182].

The relative importance between intra- and inter-layer connectivity determines most of the structural and dynamical properties of a multilayer network, which can act either as system which is structurally decoupled – i.e., consisting of independent entities – or as as a interdependent system. The transition between these two regimes for general topologies is an active research field, although in some cases it is possible to identify the existence of a sharp structural change [193, 221].

In **muxViz** it is possible to measure the correlation between layers in several ways, including [10, 129, 152, 247]:

► **Code snippet 3.1**
`layer-layer_corr.R`

- **Mean node overlapping**: measure of the number of nodes existing simultaneously in a pair of layers α and β by

$$o_n(\alpha,\beta) = m[e(W^i_j(\alpha)), e(W^i_j(\beta))]u_i/N, \tag{3.1}$$

  where $e(\cdot)$ is a function returning a vector in the space of nodes whose entries are either 0 or 1 depending if the corresponding node exists or not in a layer, while $m(\cdot)$ is a function returning the entrywise minimum of two tensors;
- **Mean edge overlapping**: measure of the number of relationships replicated across a pair of layers α and β by

$$o_e(\alpha,\beta) = \frac{2U^j_i m[W^i_j(\alpha), W^i_j(\beta)]}{W^i_j(\alpha)U^j_i + W^i_j(\beta)U^j_i}; \tag{3.2}$$

- **Inter-layer assortativity (Pearson correlation)**: measure the Pearson correlation between the degree vectors of a pair of layers, i.e., the average degree-degree correlations across layers. If $k^i(\alpha)$ indicate the (in-, out- or total) degree vector of layer α, then this is measured by

$$r_P(\alpha,\beta) = \frac{\mathrm{cov}[k^i(\alpha), k^i(\beta)]}{\sigma[k^i(\alpha)]\sigma[k^i(\beta)]}, \tag{3.3}$$

Figure 3.3: Example of multiplex network consisting of 5 layers with increasing edge overlapping. Each layer consists of 100 nodes: layer 1 is obtained from a Barabasi-Albert model, while layers 2–5 are obtained from reshuffling its connectivity while keeping an overlapping fraction of links equal to 25%, 50%, 75% and 95%, respectively.

where $\mathrm{cov}(\cdot,\cdot)$ indicates the covariance and $\sigma(\cdot)$ indicates the standard deviation;

- **Inter-layer assortativity (Spearman correlation)**: same as the previous one, but using Spearman's ρ instead of Pearson correlation[13];
- **Inter-layer similarity (by shortest-path distance between nodes)**: measure the similarity across layers of routes between nodes according to their shortest path distance. If the entries of $D^i_j(\alpha)$ indicate the shortest-path distance between any pair of nodes in layer α, and $\Delta^i_j(\alpha,\beta) = D^i_j(\alpha) - D^i_j(\beta)$, the measure is defined by

$$r_{sp} = \sqrt{\Delta^i_j(\alpha,\beta)\Delta^j_i(\alpha,\beta)}. \tag{3.4}$$

[13] Let us recall here that ρ is defined as the Pearson coefficient of the ranks rather than the value of the variables.

▸ **Code snippet 3.2**
`example_overlapping_generator.R`

The measures based on assortativity allow for a variety of options, depending if the network is directed or not. In fact, for while the total degree (T), the in-coming (I) degree and the out-going (O) degree of a node does not change in an undirected network, this is no more the case for directed relationships and, often, might be interesting to explore layer-layer correlations by considering different combinations, such as I–I, I–O, O–I and O–O, as well as T–T. Each combination allows one to gain different insights about the coupling between layers. For instance, it might be the case that hubs in one layer are usually more peripherals in another layer, in a statistically significant way. In the case of directed networks, such as many online social networks, it is possible to assess the tendency of influencers on Twitter – individuals with a high number of followers – to be influencers or not in another social network such as Instagram. This type of analysis can reveal rich structural information about the system and its actors. An example is shown in Fig. *3.2*, where a real social system is considered. While the correlation between pairs of layers is qualitatively the same across measures, the case of I–O assortativity is different: interestingly, this result is telling us that users who reply a lot to other users are more likely to be endorsed by someone else through retweets, at least for what concerns this specific data set.

We show in Fig. *3.3* an example of multiplex network consisting of five layers for increasing edge overlapping between layers. The starting network is the first layer, a Barabasi-Albert network consisting of 100 nodes: the successive layers are obtained from reshuffling its connectivity while keeping a fixed fraction of edges overlapping, respectively to 25%, 50%, 75% and 95%.

▸ **Code snippet 3.3**
`example_configmodel_generator.R`

Another example concerns the generation of multilayer networks which can be used to validate the existence of significant intra- and inter-layer correlations. This type of models can be considered the multilayer counterpart of the well-known configuration model used for the analysis of single-layer networks, with the difference that here we have two classes of models:

- **Type-I**: generate random multilayer networks with degree sequence fixed by the original one, destroying intra-layer correlations while keeping inter-layer ones;
- **Type-II**: same as Type-I but destroying inter-layer correlations too.

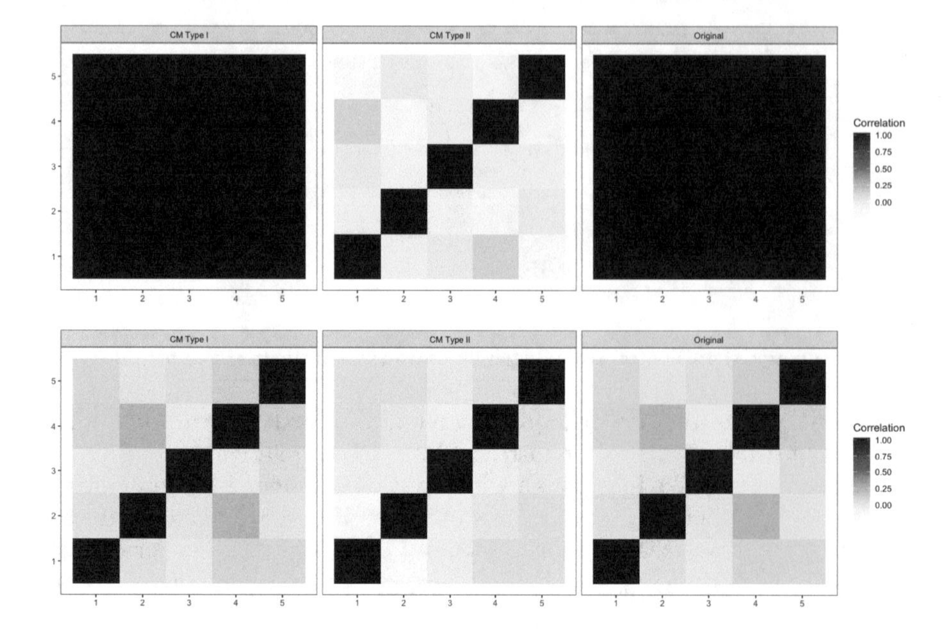

Figure 3.4: Layer-layer correlations (TT assortativity, Pearson) for multiplex networks consisting of 100 nodes and 5 layers (right-hand side column) and their Type-I (left-hand side) and Type-II (middle) configuration models (CM). In the top panels we have used the same model considered in Fig. *3.3*, whereas in the bottom panels we have used ER networks with the same wiring probability $p = 0.07$. As expected, the correlation matrices for the original network and its Type-I CM are the same, whereas the one for Type-II CM changes.

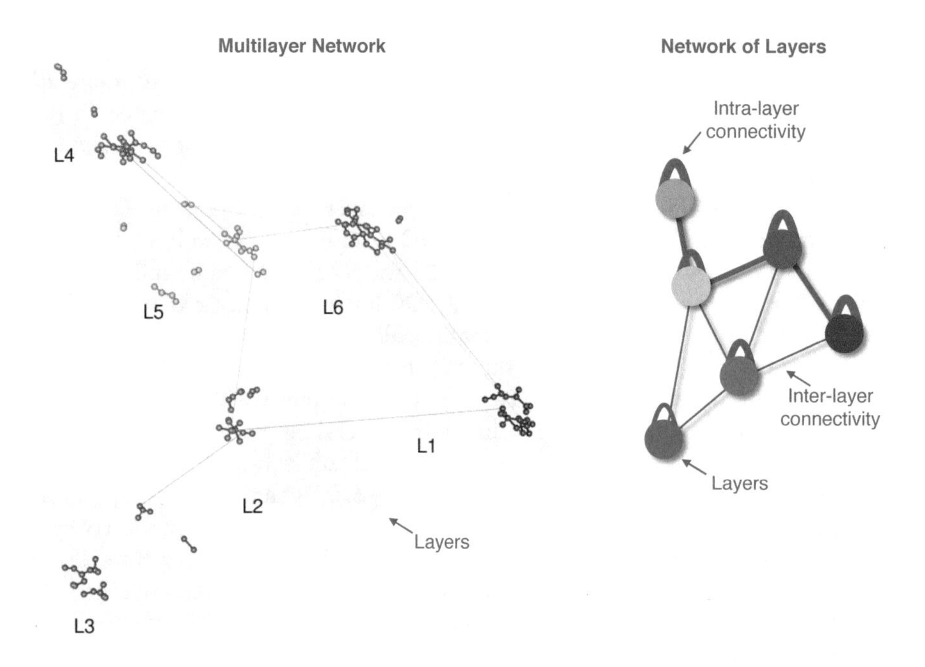

Figure 3.5: A multilayer network consisting of 6 interdependent layers and its aggregated representation showing how layers are interconnected with each other. In the latter, the thickness of the links is proportional to the number of intra- or inter-layer links they aggregate. Figure from [4] under Creative Commons Attribution-ShareAlike 4.0 International License.

Figure *3.4* shows the layer-layer correlations for two distinct multiplex networks and their corresponding multilayer configuration models. A more interactive analysis is available with the accompanying code snippet, where it is possible to inspect also the values of intra-layer correlations, not only inter-layer ones.

3.4 Network of layers

We have seen that one way to represent a multilayer network is by means of its aggregation to a classical network, where the number of nodes is preserved and edges are given by some rule (e.g, sum, average, union, intersection). However, this is not the only way to aggregate multilayer networks to obtain a coarse-grained representation of their structure. Another approach is to project the multilayer adjacency tensor in the space of layers: in the the resulting network, nodes corresponds to layers, edges encode inter-layer connectivity and self-loops encode intra-layer connectivity [129] (see Fig. ***3.5***).

Usually, this network of layers is weighted and it might be directed if inter-layer connectivity is directional. The result is a rank–2 adjacency tensor with dimensions $L \times L$, which can be mathematically obtained by

$$\Psi^{\gamma}_{\delta} = M^{i\gamma}_{j\delta} U^{j}_{i} \tag{3.5}$$

3.5 Multilayer walks, trails, paths, cycles and circuits

As in classical networks, we can define some fundamental concepts related to the exploration of a multilayer system. More specifically, let us start from the concept of ***walk***: it is defined as a sequence of adjacent nodes and edges visited by a hypothetical walker, without special constraints. Figure ***3.6*** shows an illustration of walk on a multilayer network. The *length of the walk* is generally defined by the number of traversed edges.

For classical unweighted networks, represented by binary rank–2 tensors (i.e., adjacency matrices whose entries can get only two possible values: either 0 or 1), it is possible to calculate analytically the number of walks of length ℓ from a node i to any other node j. If $i \neq j$, the walk is named *open*, whereas in the case $j = i$, the calculation gives the number of *closed* walks of length ℓ that node i is part of. If we indicate with $\mathcal{W}^{i}_{j}(\ell)$ the rank–2 tensor encoding information about walk length between any pair of nodes in the network, it is possible to show that

$$\mathcal{W}^{i}_{j}(\ell) = (A^{i}_{j})^{\ell} = A^{i}_{j_1} A^{j_1}_{j_2} \dots A^{j_{\ell-1}}_{j} \tag{3.6}$$

i.e., information can be directly obtained by calculating the entries of the ℓ–th power of the rank–2 adjacency tensor representing the network. If the links are weighted, it is still possible to use the same formalism by defining the weight of a walk as the product of the weights of the traversed links. The entries of $\mathcal{W}^{i}_{j}(\ell)$ will give the sum of weights of the walks of length ℓ connecting the corresponding pair of nodes.

It is possible to use a similar approach to calculate the same information for more complex structures such as multilayer networks. If $M^{i\alpha}_{j\beta}$ is the rank–4 multilayer adjacency tensor representing the system, then the entries of the ℓ–th power of this tensor provides the number of multilayer walks of length ℓ between a node i in layer α and a node j in layer β:

$$\mathcal{W}^{i\alpha}_{j\beta}(\ell) = M^{i\alpha}_{j_1\beta_1} M^{j_1\beta_1}_{j_2\beta_2} \dots M^{j_{\ell-1}\beta_{\ell-1}}_{j\beta}. \tag{3.7}$$

This formalism turns out to be extremely useful to highlight the topological difference between interconnected networks and their aggregated representations [129, 248]. To show this here, let $\bar{G}^{i}_{j} = M^{i\alpha}_{j\beta} U^{\beta}_{\alpha}$ be the aggregate network which accounts for inter-layer links: the corresponding rank–2 walk tensor, then

$$\begin{aligned} \bar{\mathcal{W}}^{i}_{j}(\ell) &= (\bar{G}^{i}_{j})^{\ell} = \bar{G}^{i}_{j_1} \bar{G}^{j_1}_{j_2} \dots \bar{G}^{j_{\ell-1}}_{j} \\ &= M^{i\alpha}_{j_1\beta_1} U^{\beta_1}_{\alpha} M^{j_1\beta_1}_{j_2\beta_2} U^{\beta_2}_{\beta_1} \dots M^{j_{\ell-1}\beta_{\ell-1}}_{j\beta} U^{\beta}_{\beta_{\ell-1}} \\ &= \underbrace{\left(M^{i\alpha}_{j_1\beta_1} M^{j_1\beta_1}_{j_2\beta_2} \dots M^{j_{\ell-1}\beta_{\ell-1}}_{j\beta}\right)}_{\mathcal{W}^{i\alpha}_{j\beta}(\ell)} \underbrace{\left(U^{\beta_1}_{\alpha} U^{\beta_2}_{\beta_1} \dots U^{\beta}_{\beta_{\ell-1}}\right)}_{U^{\alpha}_{\beta} L^{\ell-1}}, \end{aligned} \tag{3.8}$$

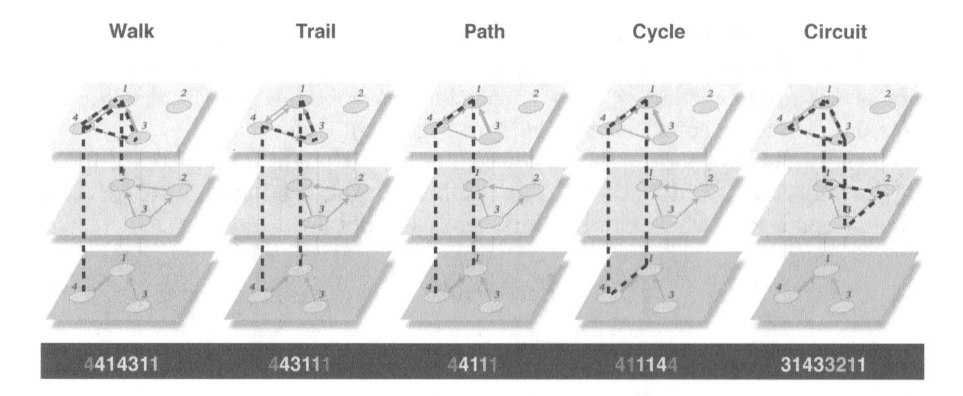

Figure 3.6: Different types of walks on a multilayer network. A multilayer walk is the most general way to traverse nodes and links of a multilayer system. One can apply some restrictions on the number of repeated nodes or links, as well as the identity of origin and destination nodes, to define special types of walks, such as multilayer trails, paths, cycles and circuits. In the illustration, a sequence of nodes and edges corresponding to each type of walk is shown with a dashed line, with the corresponding sequence of visited nodes reported in the bottom line. See the text for further details. Figure from [4] under Creative Commons Attribution-ShareAlike 4.0 International License.

showing that the number of walks of length ℓ between two nodes in the aggregate network is not a linear function of the number of walks of length ℓ between the same pair of nodes in the multilayer network.

On can apply some restrictions to a multilayer walk, obtaining more peculiar ways of traversing edges. For instance, one might be interested in walks where links can be traversed only one time: this restriction on repeated links defines a ***multilayer trail*** (Fig. ***3.6***). An open multilayer trail where one further applies the restriction that repeated nodes are not allowed defines a ***multilayer path*** (Fig. ***3.6***), whereas a closed trail where only the origin and destination nodes are repeated, thus closing the walk, is named a ***multilayer cycle*** (Fig. ***3.6***). If a closed trail allows for more than one repeated node, then we have a ***multilayer circuit*** (Fig. ***3.6***).

Shortest paths are among the most important walks in a network: they allow to model, for instance, how information is exchanged between two nodes by using the smallest number of hops, i.e., the least number of traversed nodes and links. Among all the paths between two nodes in a multilayer network, the shortest path is the shortest one and its length is traditionally used to define a distance between nodes. This notion of *geodesic distance* is valid if the network is undirected, because the triangular inequality – one of the three properties defining a metric distance – is no more satisfied in the case of directed links[14].

[14] Multilayer shortest paths and descriptors depending on it are available in the `muxViz` LIB but not (yet) in the GUI.

Chapter 4 | Multilayer Versatility and Triads

Contents

THE ability to identify nodes playing a special role in a complex system can be crucial for a wide spectrum of applications, from anticipating urban areas going to be congested by traffic flow to maximizing the individuals' engagement during marketing campaigns or devising more effective containment strategies against the spreading of infectious agent within structured populations.

As in the case of classical networks, it is possible to define multiple notions of node *importance*, *influence* or *relevance*, depending on the application of interest. In general, there are two classes of centrality measures: the ones developed for non-interconnected multiplex systems and the ones developed for interconnected multilayer topologies. Despite several multiplex centrality measures have been recently proposed [151, 175, 249–253], here we mostly focus on the ones developed for the second class and included in `muxViz`, which are naturally derived from their classical counterparts by exploiting the tensorial framework [10, 129, 162, 254] and are better known as *versatility measures.*

▸ **Code snippet 4.1**
`versatility_measures.R`

4.1 Node centrality in multilayer networks

One might wonder to which extent the calculation of versatilities, in general more expensive than their classical counterparts, is providing any insight with respect to:

- calculating the same measure on each layer separately and then aggregate the results according to some heuristics;
- calculating the same measure on the aggregate network.

The short answer is that considering the layers of coupled systems in isolation or aggregate them to a classical network might be a provide a very poor model of network structure and dynamics. The long answer is that in the first scenario the way one aggregates the results might significantly alter them and, overall, one is completely neglecting the possibly existing structural correlations which are usually expected to play a crucial role to define the importance of each node. In the second scenario, one washes out existing correlations and introduces spurious paths through which information diffuses, altering the estimation of node centrality. Moreover, aggregation might introduce a degeneration in local structures which hinder the identification of central nodes.

An illustration is shown in Fig. *4.1*, where a system consisting of scientists (nodes) collaborating (edges) to produce a scientific paper is considered [10]. Since a scientific study is the result of several interdependent tasks, from conceiving the study itself to make experiments and write the paper, each task can

M. De Domenico, *Multilayer Networks: Analysis and Visualization*, https://doi.org/10.1007/978-3-030-75718-2_4

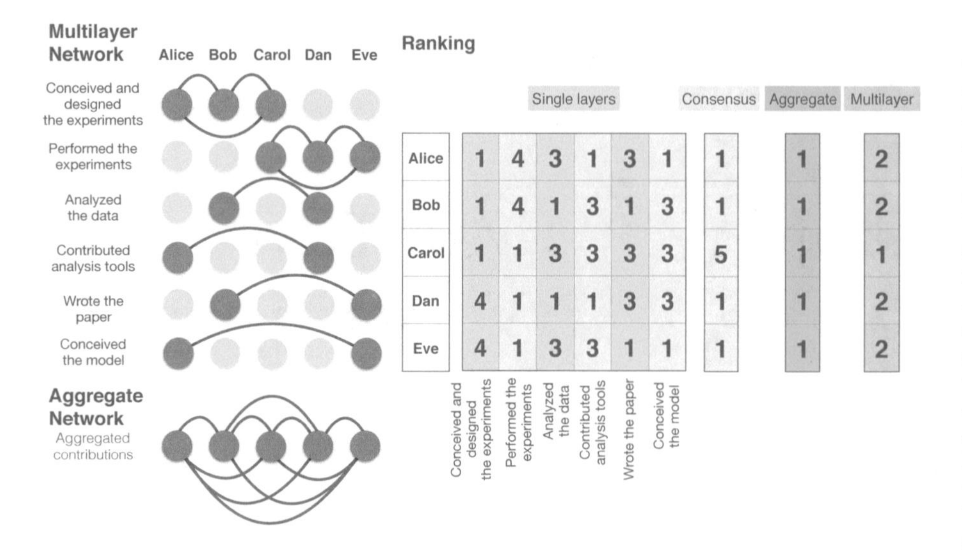

Figure 4.1: Illustrative example of a multiplex network of authors contributing to a research article. Five authors collaborate in different tasks (e.g. "Analyzed the data" or "Wrote the paper") – defining distinct layers – and are connected by one edge if they work to the same task. The aggregate network is also shown: it corresponds to a clique. Any centrality analysis of the aggregate representation is not meaningful, because all authors will be central in the same way due to the symmetry of the topology. The analysis of layers in isolation produces a ranking for each task: the 6 different rankings must be heuristically aggregated into a single ranking, for instance by using a consensus rule. The result in this case suggests all authors have the same importance, except Carol who is the less central one: she contributed to two tasks instead of three, as the other authors, therefore she is under-represented and under-ranked by the heuristics. However, Carol is the only author active in the two tasks with the largest number of active authors, playing a crucial role for the exchange of information between two non-overlapping groups of authors: the multilayer analysis, in fact, ranks Carol as the most *versatile* author, in agreement with expectations [10]. Figure from [4] under Creative Commons Attribution-ShareAlike 4.0 International License.

be considered as a distinct layer. The specific example shows how applying a reasonable heuristics – such as consensus – to the results obtained from each layer separately or calculating centrality from the aggregate representation of the system leads to misleading results. In fact, single-layer analysis misses to identify the cross-layer role of Carol, while aggregate analysis is useless due to the degeneration of the topology into a clique, i.e., a graph where all nodes are connected with each other and none of them is more important than the others.

In the following, we will refer to *multilayer centrality* or *versatility* to indicate the same concept.

4.1.1 Multilayer degree and strength centralities

The easiest centrality to measure is the degree. As for its classical counterpart, the multilayer degree is a local measure quantifying the number of edges incident to a node [129]. If the network is directed, we usually distinguish between *in-degree* – counting edges directed to a node from its neighbors – and *out-degree* – counting edges directed from a node to its neighbors. The total degree is the sum of the two degrees when the network is directed, whereas their half-sum is used when the network is undirected to avoid double counting edges. The multilayer degree is usually defined in terms of operation on the multilayer adjacency tensor: however, in general this tensor can represent weighted networks and to avoid ambiguity in the following definitions let us first introduce the function $\mathcal{B}(\cdot)$ which act on each entry of a tensor and returns 1 if the entry is larger than zero, and zero otherwise. In fact, $\mathcal{B}(\cdot)$ is a function that binarizes a tensor.

There are also two more cases to distinguish, depending on wether or not one is interested in accounting for inter-layer connectivity or not. If inter-layer edges should not be accounted for, then the multilayer in-degree is defined by

$$k^i = \mathcal{B}(M^{i\alpha}_{j\alpha})u^j = \mathcal{B}(G^i_j)u^j, \tag{4.1}$$

whereas the multilayer out-degree vector is defined by

$$k_j = \mathcal{B}(M^{i\alpha}_{j\alpha})u_i = \mathcal{B}(G^i_j)u_i. \tag{4.2}$$

The careful reader has noticed that the difference between the two definition is based on the 1–tensor used for contracting the multilayer adjacency tensor: in-degree is represented by a contravariant vector, whereas out-degree is represented by a covariant vector. The difference between the two measures vanishes if the multilayer network is undirected and both of them coincide with the total degree. Moreover, we have easily shown that this type of multilayer degree coincides with the degree calculated from the corresponding binarized aggregate network.

If inter-layer edges should be accounted for, then the multilayer in-degree is defined by

$$K^i = \mathcal{B}(M^{i\alpha}_{j\beta})U^{\beta}_{\alpha}u^j = \mathcal{B}(\bar{G})^i_j u^j \tag{4.3}$$

whereas the multilayer out-degree vector is defined by

$$K_j = \mathcal{B}(M^{i\alpha}_{j\beta})U^{\beta}_{\alpha}u_i = \mathcal{B}(\bar{G})^i_j u_i. \tag{4.4}$$

In practice, it corresponds to first project the multilayer network into its aggregate network, while adding self-loops to nodes existing in multiple layers.

However, in many applications the information about edge weights can be important and another set of measures is usually used: multilayer strength centralities. The definitions are formally identical to the ones defining the degree, where $M^{i\alpha}_{j\beta}$ is used instead of $\mathcal{B}(M^{i\alpha}_{j\beta})$. If inter-layer connectivity should not be taken into account, then multilayer in-strength and out-strength centralities are defined by

$$s^i = M^{i\alpha}_{j\alpha}u^j = G^i_j u^j \tag{4.5}$$

$$s_j = M^{i\alpha}_{j\alpha}u_i = G^i_j u_i, \tag{4.6}$$

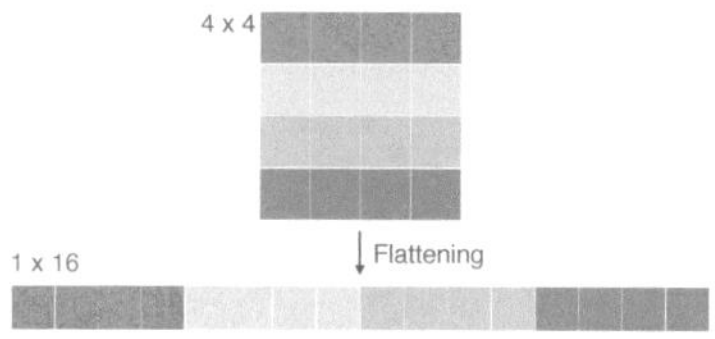

respectively. Conversely, if inter-layer edges should be accounted for, then the multilayer in-strength and out-strength centralities are defined by

$$S^i = M^{i\alpha}_{j\beta}U^{\beta}_{\alpha}u^j = \bar{G}^i_j u^j \tag{4.7}$$

whereas the multilayer out-strength vector is defined by

$$S_j = M^{i\alpha}_{j\beta}U^{\beta}_{\alpha}u_i = \bar{G}^i_j u_i. \tag{4.8}$$

In this second case, it is worth noticing that the calculation corresponds to the one made from the corresponding aggregate network, where self-loops are added to nodes existing in multiple layers with a weigh equal to the sum of the corresponding inter-layer weights.

4 x 4

Flattening

1 x 16

Figure 4.2: Flattening a 4 × 4 rank–2 tensor into a 1 × 16 rank–1 tensor, while preserving information. Figure from [4] under Creative Commons Attribution-ShareAlike 4.0 International License.

4.1.1.1 Multilayer eigenvector centrality

A widely used measure of importance in classical networks is eigenvector centrality, whose calculation is based on finding the leading eigenvector v_i of the rank–2 adjacency tensor:

$$W^i_j v_i = \lambda v_j. \tag{4.9}$$

However, in the case of multilayer networks the same problem is extended to rank-4 tensors, leading to the problem of finding the leading *eigentensor* from the equation

$$M^{i\alpha}_{j\beta}V_{i\alpha} = \lambda_1 V_{j\beta}, \tag{4.10}$$

which in general is a hard problem and the solution could not be unique. One possible largely adopted approach to solve this problem is based on *matricization* – or *flattening* – of $M^{i\alpha}_{j\beta}$, corresponding to its unfolding to lower rank

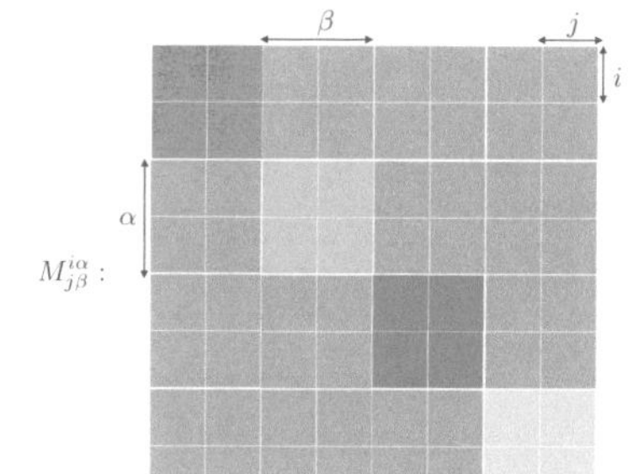

Figure 4.3: Result of flattening a multilayer adjacency tensor $M^{i\alpha}_{j\beta}$, representing a system with $N = 2$ nodes and $L = 4$ layers into a $(NL) \times (NL)$ supra-adjacency matrix, while preserving information. Figure from [4] under Creative Commons Attribution-ShareAlike 4.0 International License.

tensors [7]. In fact, we have already discussed in the previous chapter about the result of this operation: the supra-adjacency matrix (Fig. ***1.13***).

The flattening can transform, for instance, a $N \times N$ rank–2 tensor such as W^i_j into a rank–1 tensor w^k with N^2 components (Fig. ***4.2*** and ***4.3***). Similarly, a rank–4 $N \times N \times L \times L$ tensor such as $M^{i\alpha}_{j\beta}$ can be unfolded into a squared rank–2 tensor like the supra-adjacency matrix $\tilde{M}^k_l$ with $NL \times NL$ components[15].

[15] There are as many representations of this flattening as the number of permutations of diagonal blocks of size N^2, i.e., $L!$. Unfolding does not alter the spectral properties, but one should be very careful when using the output of algorithms.

The unfolded tensors can be used to solve the original eigenvalue problem for rank–4 tensors. The corresponding leading eigenvector $\tilde{v}_l$ is the solution of the equation

$$\tilde{M}^k_l \tilde{v}_k = \tilde{\lambda}_1 \tilde{v}_l, \tag{4.11}$$

which is a *supra-vector* with NL components corresponding to the flattening of the eigentensor $V_{i\alpha}$. This corresponds to calculate the well known Bonacich's eigenvector centrality of each node in each layer, while accounting for the whole interconnected structure of the layers.

However, the solution is not always compatible with one's expectation: in fact, one is often interested in quantifying the centrality of each node with a single number, rather than a whole vector. If the contribution of each layer is equivalent, the same weight can be assigned to all layers and the overall eigenvector versatility is simply obtained by summing up the corresponding scores across layers as

$$v_i = V_{i\alpha} u^\alpha, \tag{4.12}$$

an approach validated by the calculation of the same score by means of agent-based simulations [10].

More generally, if layers should not be considered equivalent and one can reasonably assign different weights, a vector ω^α can be used instead, with $\omega^\alpha = u^\alpha$ in the case of equally contributing layers. The measure is usually normalized for practical purposes and to allow for the comparison across different networks, for instance against results obtained from single-layer and aggregate network analysis.

Box 4.1.1: Comparing versatility against their classical counterpart

The eigenvector centrality calculated from the p–th layer is given by

$$\begin{aligned} v_j(p) &= \lambda_1^{-1}(p) G^i_j(p) v_i(p) \\ &= \lambda_1^{-1}(p) M^{i\alpha}_{j\beta} E^\beta_\alpha(pp) v_i(p). \end{aligned} \tag{4.13}$$

To obtain an overall centrality vector, an heuristics for the aggregation of results must be used and one of the simplest choices might be to sum up over layers:

$$\begin{aligned} \tilde{v}_j &= \sum_{p=1}^{L} \lambda_1^{-1}(p) M^{i\alpha}_{j\beta} E^\beta_\alpha(pp) v_i(p) \\ &= M^{i\alpha}_{j\beta} \sum_{p=1}^{L} \lambda_1^{-1}(p) E^\beta_\alpha(pp) v_i(p) \end{aligned} \tag{4.14}$$

Similarly, the eigenvector centrality obtained from the aggregate network (accounting for inter-layer edges) is given by

$$\begin{aligned} \bar{v}_j &= \bar{\lambda}_1^{-1} \bar{G}^i_j \bar{v}_i \\ &= \bar{\lambda}_1^{-1} M^{i\alpha}_{j\beta} u^\beta, \qquad \bar{V}_{i\alpha} = \bar{v}_i u_\alpha \end{aligned} \tag{4.15}$$

whereas the multilayer eigenvector centrality is obtained as

$$v_j = \lambda_1^{-1} M^{i\alpha}_{j\beta} V_{i\alpha} u^\beta. \tag{4.16}$$

The difference between the multilayer measures and its classical counterparts are given by

$$v_j - \bar{v}_j = M^{i\alpha}_{j\beta} u^{\beta} [\lambda_1^{-1} V_{i\alpha} - \bar{\lambda}_1^{-1} \bar{V}_{i\alpha}], \qquad \bar{V}_{i\alpha} = \bar{v}_i u_{\alpha}, \tag{4.17}$$

and

$$v_j - \bar{v}_j = M^{i\alpha}_{j\beta} [\lambda_1^{-1} V_{i\alpha} u^{\beta} - \sum_{p=1}^{L} \lambda_1^{-1}(p) E^{\beta}_{\alpha}(pp) v_i(p)]. \tag{4.18}$$

The above formulas show that the two vectors are non-trivially related with each other. It is possible to perform the analysis of how their difference varies with inter-layer coupling, for instance by using eigenvalue perturbation analysis.

4.1.2 Multilayer Katz centrality

Katz originally introduced this centrality descriptor to overcome some limitations of eigenvector centrality, especially in applications to directed networks. In the context of multilayer analysis, the Katz versatility is obtained by solving the tensorial equation [10]

$$\Phi_{j\beta} = a M^{i\alpha}_{j\beta} \Phi_{i\alpha} + b u_{j\beta}, \tag{4.19}$$

given by

$$\Phi_{j\beta} = [(\delta - aM)^{-1}]^{i\alpha}_{j\beta} U_{i\alpha}, \tag{4.20}$$

where $\delta^{i\alpha}_{j\beta} = \delta^{i}_{j} \delta^{\alpha}_{\beta}$, a is a constant smaller than the largest eigenvalue and b is another constant generally equal to 1. As for the eigenvector versatility, the Katz versatility across layers is obtained by an appropriate contraction with the 1–tensor:

$$\phi_i = \Phi_{i\alpha} u^{\alpha}, \tag{4.21}$$

or, more generally, with a vector ω^{α} accounting for the weight of each layer.

4.1.3 Multilayer HITS centrality

The Hyperlink-Induced Topic Search (HITS) centrality introduced by Jon Kleinberg, in its original conception, provided a network-based approach to rank pages of the World Wide Web, released by the European Organization for Nuclear Research (CERN) in 1991. Like the Katz centrality, this centrality overcomes some limitations of the eigenvector centrality and can be safely applied to directed networks. The method distinguishes two classes of nodes: *authorities*, which are nodes pointed by *hubs*, which are nodes pointing many other nodes.

The multilayer versatility corresponding to HITS centrality [10] can be obtained by solving, simultaneously, the two eigenvalue problems

$$(MM^{\top})^{i\alpha}_{j\beta} \Gamma_{i\alpha} = \lambda_1 \Gamma_{j\beta} \tag{4.22}$$

$$(M^{\top}M)^{i\alpha}_{j\beta} \Upsilon_{i\alpha} = \lambda_1 \Upsilon_{j\beta}, \tag{4.23}$$

where $\top$ denotes the transpose operator. The two solutions, $\Gamma_{i\alpha}$ and $\Upsilon_{i\alpha}$, quantify hub and authority versatility, respectively. As for the other eigenvector-based versatilities previously considered, these two descriptors provide a score per node and per layer which are usually combined as

$$\gamma_i = \Gamma_{i\alpha} u^\alpha \tag{4.24}$$

$$\upsilon_i = \Upsilon_{i\alpha} u^\alpha, \tag{4.25}$$

to provide a unique overall versatility vector. More generally, the contraction can be performed with a vector ω^α accounting for the weight of each layer.

4.1.4 Multilayer PageRank centrality

At the end of the '90s the problem of ranking Web pages was crucial: the rapid growth of the World Wide Web made the search for specific information a technological challenge, potentially solvable by ranking web pages by their importance with respect to how information flows in the network. Two young students enrolled in Stanford University, Sergey Brin and Larry Page, had the great intuition to overcome limitations of eigenvector and Katz centralities by considering a random walker which, once in a node (a Web page) can either jump to another Web page in its neighborhood or make a teleportation anywhere in the network – even to disconnected components –, the latter action corresponding to moving to a Web page chosen randomly with uniform probability [255]. The new algorithm, named PageRank, is one of the most powerful method for ranking and it is the core of how Google Inc. (now Google LLC) discovers the most relevant Web pages out of billions ones.

The natural extension of the original method to multilayer networks [10] is based on multilayer random walks [18] (see Fig. *2.4*) and accounts for the interplay between *jumping* on nodes in the local neighborhood, within the same layer, and *switching* layer through inter-layer edges. Other variants, based on specific prescriptions for coupling layers when inter-layer connectivity is not available, have been proposed [250, 251].

Here, we focus on PageRank versatility, which can be obtained as the steady-state solution of a special Markov process on the multilayer network. Random walkers explore the network according to a special multilayer transition tensor and their dynamics is governed by a master equation identical to Eq. (2.6), whose solution is formally equivalent to the leading eigenvector of the transition tensor. In the case of interconnected multilayer networks, this tensor is given by

$$R^{i\alpha}_{j\beta} = rT^{i\alpha}_{j\beta} + \frac{(1-r)}{NL} u^{i\alpha}_{j\beta}, \tag{4.26}$$

with r a constant, generally set to 0.85 like in the original Google algorithm [255], N is the number of nodes per layer and L is the number of layers. Here, $R^{i\alpha}_{j\beta}$ governs the dynamics of a random walk within and across layers and might be named *Google tensor*. Similarly to its classical counterpart, the tensor accounts for two different contributions: i) a transition to multilayer neighborhood – i.e., all the nodes reachable within one time step by structural connectivity, either intra- or inter-layer one – with rate r; ii) a teleportation with uniform probability to any node in the system, regardless of the layer, with rate $1-r$. The PageRank versatility per node and per layer $\Pi_{i\alpha}$ is obtained by solving the tensorial equation

$$R^{i\alpha}_{j\beta} \Pi_{i\alpha} = \Pi_{j\beta}, \tag{4.27}$$

which, as for the other eigenvector-based versatilities previously considered, is usually combined as

$$\pi_i = \Pi_{i\alpha} u^\alpha \tag{4.28}$$

to provide a unique overall versatility vector. More generally, the contraction can be performed with a vector ω^{α} accounting for the weight of each layer.

4.1.5 Multilayer k–coreness centrality

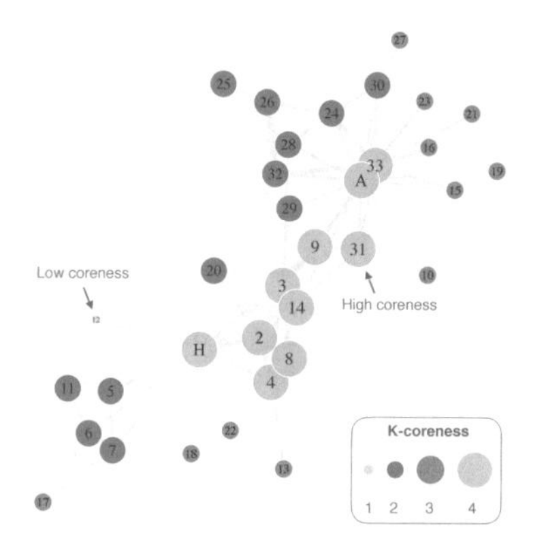

Figure 4.4: Zachary Karate Club where nodes are colored and sized by their k–coreness [11].

Often, it is useful to characterize the centrality of nodes by their relationship with tightly interconnected groups acting as *cores* for the system. One measure of this relationship is known as *coreness* [11], quantifying the centrality of nodes because they are part of a k–core, the largest group of nodes which have at least degree k within the group. For instance, a node i in a 4–core is required to have at least four edges (ie, $k_i = 4$) to *all* the nodes of that core.

The coreness of individuals in the Zachary Karate Club is shown in Fig. *4.4*, highlighting the presence of a 4–core, a 3–core and a 2–core of almost the same size. The identification of k–cores can be important to identify, for instance, influential spreaders in complex networks [122] – i.e., those actors who spread information faster and more efficiently than other ones, such as hubs – or to decompose communication networks, such as the Internet, in order to better understand their robustness [256]. K–core decomposition is especially useful in analysis of large-scale complex networks, because they allow for a meaningful mapping of the underlying structure and the characterization of many salient structural features also from visual inspection (Fig. *4.5*).

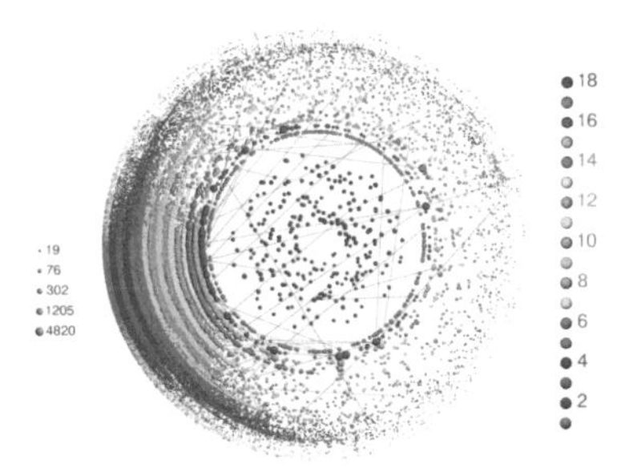

Figure 4.5: K-core decomposition of the large-scale social network activity on Twitter during the discovery of the Higgs boson. Reproduced from Ref. [12] under Creative Commons Attribution 4.0 International License http://creativecommons.org/licenses/by/4.0/

In the case of multilayer networks, the concept of k–coreness is extended to account for coreness in each layer [257]. Let $K_{i\alpha}$ indicate the coreness of each node in each layer: note that, at variance with previous calculations, this tensor is not the solution of an eigenvalue problem. For a fixed layer, e.g., layer 1, the components of K_{i1} correspond to the coreness of all nodes in that layer, calculated as explained above in the case of single-layer networks. The overall coreness of a node is obtained by taking the minimum coreness across all layers, defining the multilayer coreness versatility κ_i.

4.1.6 Multilayer Closeness centrality

▸ **Code snippet 4.2** `example_plot_edgecolored_paths.R`

▸ **Code snippet 4.3** `example_plot_edgecolored_paths_coupling.R`

Often, it can be useful to quantify how each node, on average, is "close" to all the other nodes in the network. To this aim, a distance measure must be defined first. Historically, one of the most widely used measures of distance between pair of nodes is the shortest path to connect them. Once a distance measure is defined, it is possible to define a new centrality descriptor known as *closeness* [258]. However, the original mathematical definition of closeness is not suitable for general networks, as the ones with disconnected components or isolated nodes. Tore Opshal and colleagues proposed in 2010 to define closeness centrality [259] by

$$c_i = \sum_{j=1}^{N} \frac{1}{d_{ij}}, \qquad (4.29)$$

where d_{ij} is the shortest-path distance between nodes i and j, and $d_{ij} = \infty$ if i and j belongs to disconnected components (see Sec. 5.1 for a description of multilayer connected components).

Multilayer closeness versatility[16] is defined as its classical counterpart, where multilayer shortest-path (see Sec. 3.5) distances are considered instead. The value of this versatility descriptor is 0 if a node is disconnected from all other

[16] In `muxViz` it is available only through the standalone library.

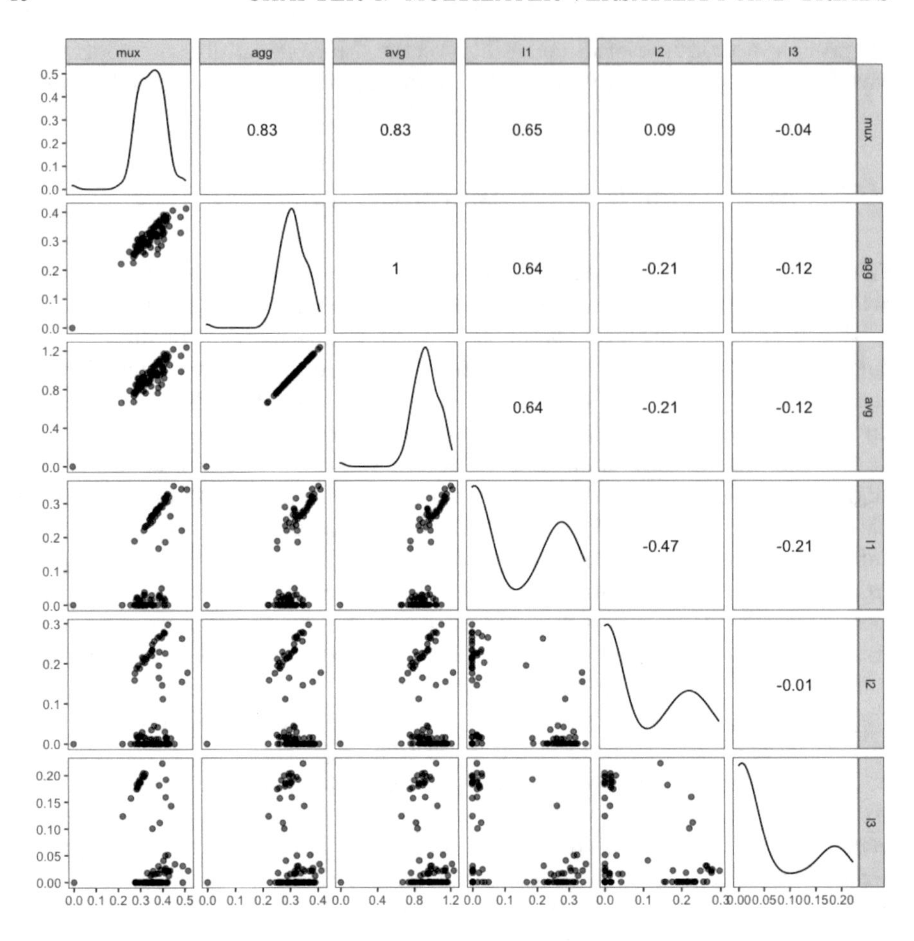

Figure 4.6: Correlation plot for closeness versatility and centralities obtained from a non-interconnected multiplex network with 3 layers. Layer 1 consists of three groups of size 20, 30 and 40 nodes respectively, while layers 2 and 3 are obtained from randomly deleting edges from the first layer, to keep some topological correlations across the whole system. Further, the edges are weighted with uniformly random numbers between 0.5 and 1, to influence path statistics.

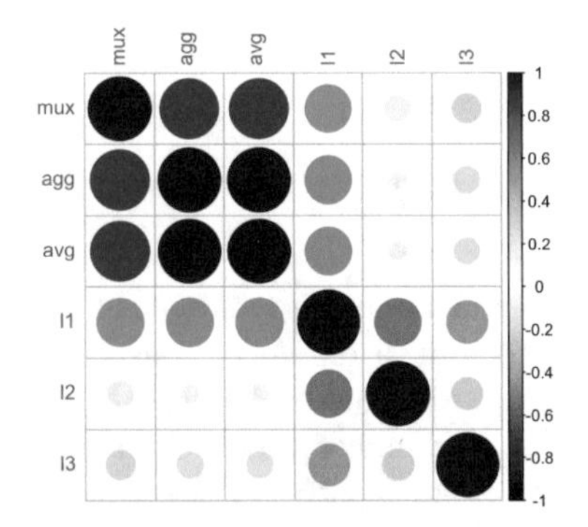

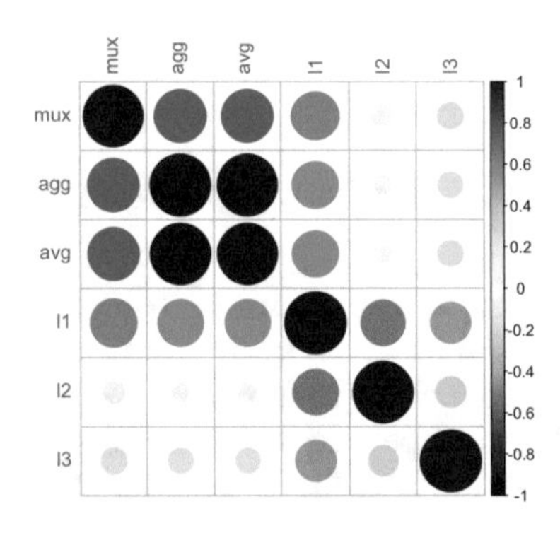

Figure 4.7: Left: as in Fig. ***4.6***, a different visualization. Right panel: the same setup used in Fig. ***4.6*** with the addition of inter-layer connectivity with a strong weight, to put in evidence the effects of inter-layer coupling on path statistics on correlations between the multiplex system – which is the only network affected by inter-layer connectivity – and the other representations.

nodes and can be normalized to get the value of 1 if a node is connected to all other nodes.

For this specific case, we consider a non-interconnected multiplex network consisting of 3 layers and calculate path statistics, including closeness, for this system, its aggregate networks (sum and average are considered) and each layer separately. The results of the analysis are shown in Fig. ***4.6*** and Fig. ***4.7*** (left panel), and are fully reproducible with the provided code snippets.

From Fig. ***4.6*** it is clear that closeness versatility correlates with closeness centrality calculated from the aggregated representations, whereas the correlation degrades with some layers. Overall, the correlation appears to be driven by the first layer where, in fact, anti-correlations are observed with the other two layers.

4.1.7 Application to fictional social networks

We show here a practical implementation of these ideas in the case of two real multiplex networks obtained from well-known fictional societies: the Game of Thrones (GoT) series and (part of) the Star Wars (SW) saga.

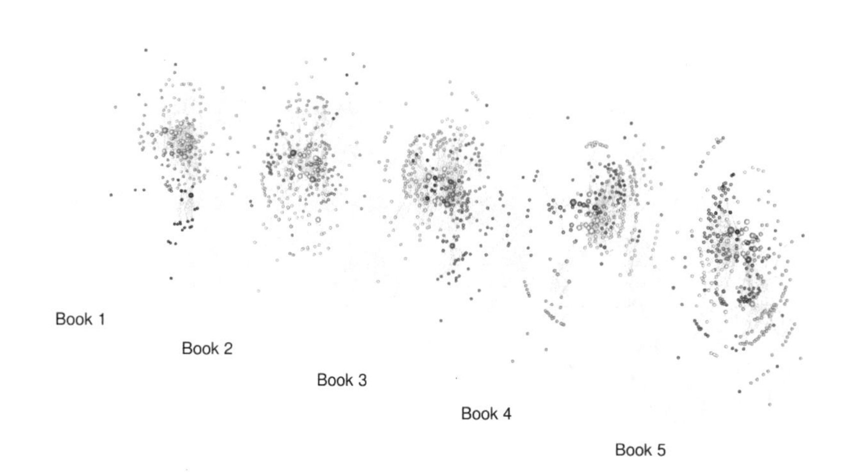

Figure 4.8: Game of Thrones multiplex network: layered visualization of the character interaction network (see the text for details), where layers encode books of George R. R. Martin's "A Song of Ice and Fire" saga and nodes are characters. Colors indicate the multilayer community membership. Figure from [4] under Creative Commons Attribution-ShareAlike 4.0 International License.

The multiplex GoT is the character interaction networks built from George R. R. Martin's "A Song of Ice and Fire" saga. Each layer corresponds to a book of the series, where a link connects two characters if their names (or nicknames) appeared within 15 words of one another in one that books. The weight of the edges corresponds to the number of interactions across the whole book.

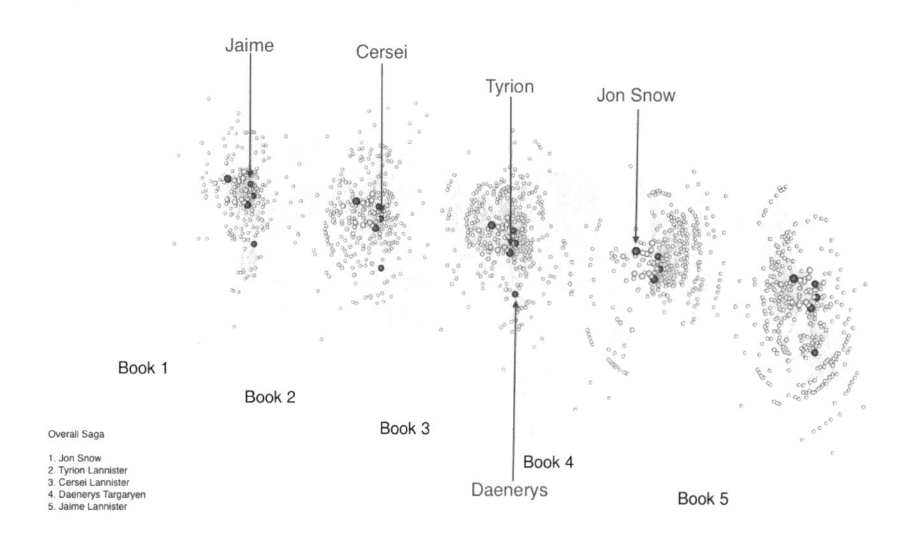

Figure 4.9: Game of Thrones multiplex network: layered visualization as in Fig. *4.8*, where nodes with the top PageRank versatility are highlighted. Figure from [4] under Creative Commons Attribution-ShareAlike 4.0 International License.

One can calculate the community structure (Fig. *4.8*) as well as PageRank versatility (Fig. *4.9*) and plot the resulting non-interconnected multiplex network by coloring nodes accordingly. Interestingly, the PageRank versatility highlights that the most influential characters are Jon Snow, Tyrion Lannister, Cersei Lannister, Daenerys Targaryen and Jaime Lannister, a result which is mostly in agreement with the author's personal view of the story.

With `muxViz` it is also possible to perform a thorough analysis to compare different types of versatility descriptors against their single-layer coun-

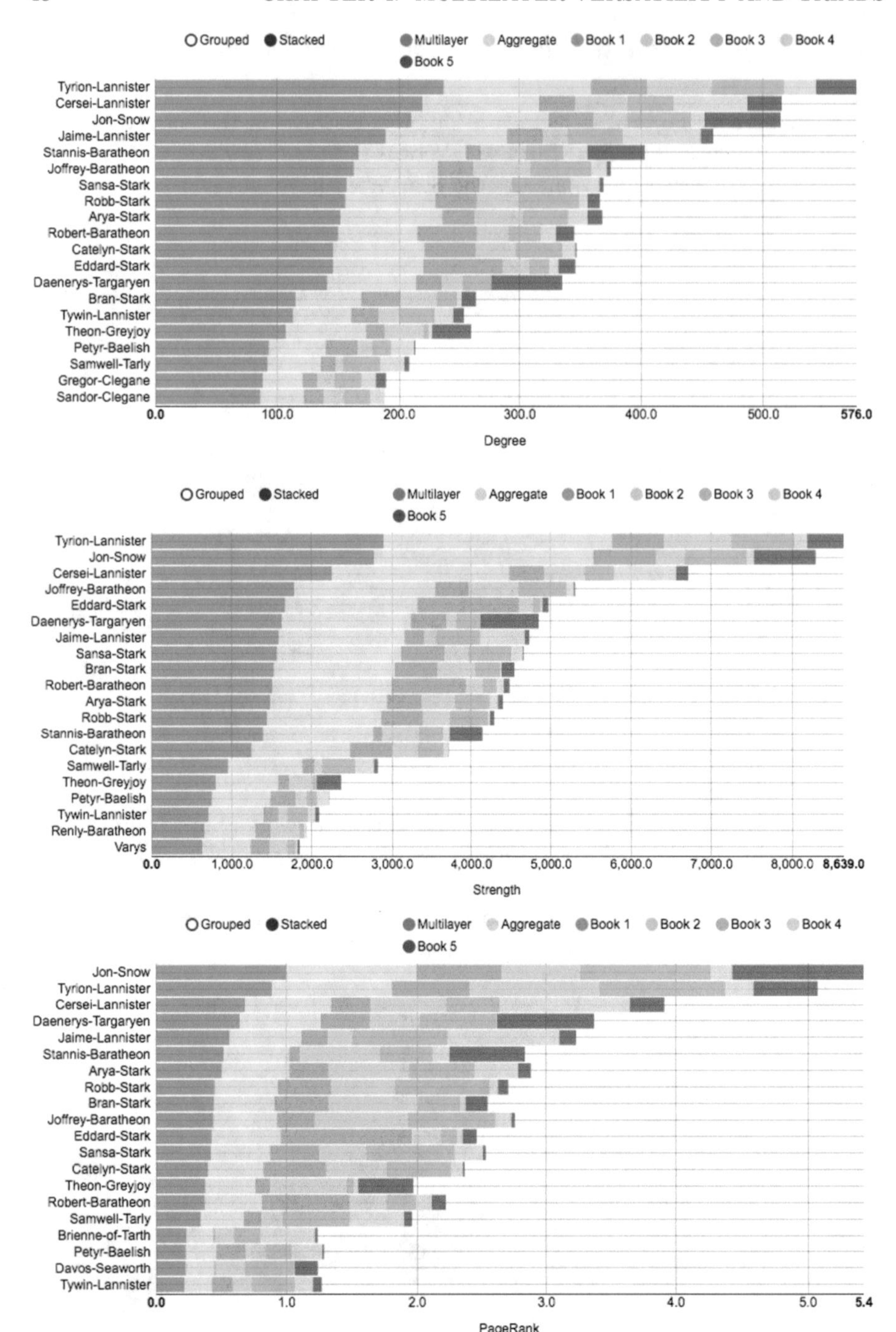

Figure 4.10: Game of Thrones multiplex network: stacked visualization of rankings obtained from degree, strength and PageRank versatility and centrality analysis, respectively. The results obtained from multilayer analysis can be easily compared against the results obtained from the aggregate and single-layer analyses.

terparts obtained from the aggregate network and from each layer separately. Figure *4.10* shows the results of this type of analysis in the case of degree, strength and PageRank versatility and centralities. Remarkably, single-layer analysis provides only a partial view of the influence of a character across the whole saga, and the aggregate network is not always a faithful representation, mixing cross-layer information in an uncontrollable way. It is also worth remarking how different measures provide different rankings: Tyrion and Cersei Lannister are the most social characters (as measured by the degree centrality) while Daenerys Targaryen is not even in the top 10; PageRank captures influence beyond the local neighborhood and Jon Snow becomes the most important character, with Daenerys Targaryen jumping to the fourth position.

Similarly, one can study the character interaction network of the SW saga, the franchise created by George Lucas. Here, layers correspond to episodes

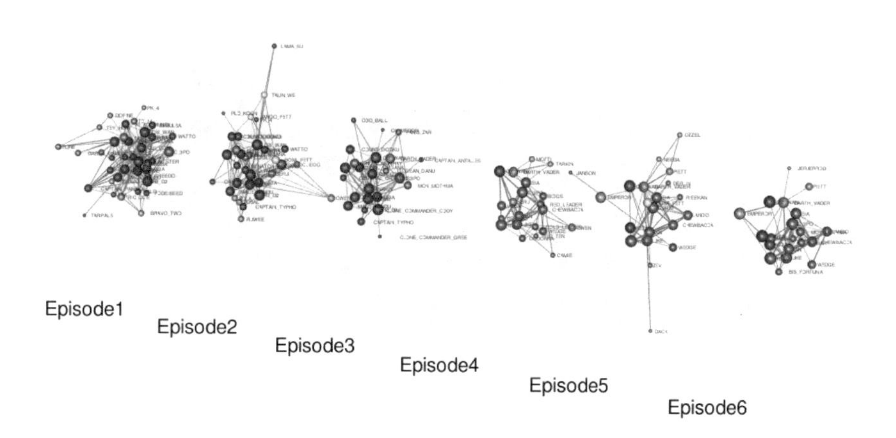

Figure 4.11: Star Wars multiplex network: layered visualization of the character interaction network (see the text for details), where layers encode episodes of George Lucas' franchise and nodes are characters. Colors indicate the multilayer community membership. Figure from [4] under Creative Commons Attribution-ShareAlike 4.0 International License.

(from first to sixth, in this example), nodes are characters and links encode who speaks to whom within the same scene of the corresponding movie transcript[17].

[17] I acknowledge the outstanding work done by Evelina Gabasova, who prepared the original data, here used under Creative Commons Attribution 4.0 International Public License. The original data set, not in `muxViz` format, is available at https://github.com/evelinag/StarWars-social-network/.

Figure *4.11* shows a layered visualization of the SW multiplex network, where nodes are colored by community and sized by their PageRank versatility, whereas Fig. *4.12* shows the comparison among PageRank versatility and PageRank centralities obtained from the aggregate and the single-layer analysis. Remarkably, the analysis of the aggregate network suggests that R2-D2 – a droid playing a very important role in the saga – is the most central character: for SW fans this might sound disappointing, despite the crucial role of the droid in many events. The multiplex analysis suggests that Anakin Skywalker is the most influential character, with Obi Wan ranked third and other crucial characters such as Luke Skywalker, Padme Amidala, Han Solo, Leia Organa and Qui-Gon Jinn in the top ten.

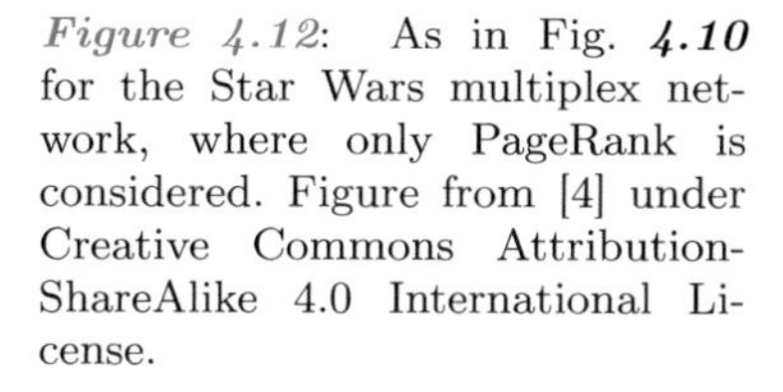

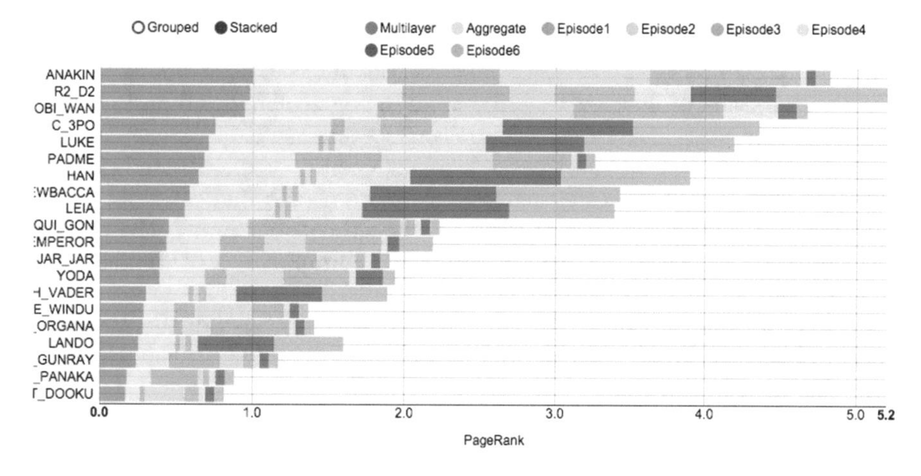

Figure 4.12: As in Fig. *4.10* for the Star Wars multiplex network, where only PageRank is considered. Figure from [4] under Creative Commons Attribution-ShareAlike 4.0 International License.

4.2 Multilayer motifs

Many complex systems consists of specific small subnetworks which are observed more frequently than their counterparts in random networks. These small subnetworks are often considered as the building blocks of complex networks and their analysis shed light on several structural and functional properties, from molecular interactions in biochemistry and synaptic connections in connectomics to food webs in ecology and World Wide Web [260–263]. Motifs involving 3 and 4 nodes are shown in Fig. *4.13* and Fig. *4.14*. Motifs are able

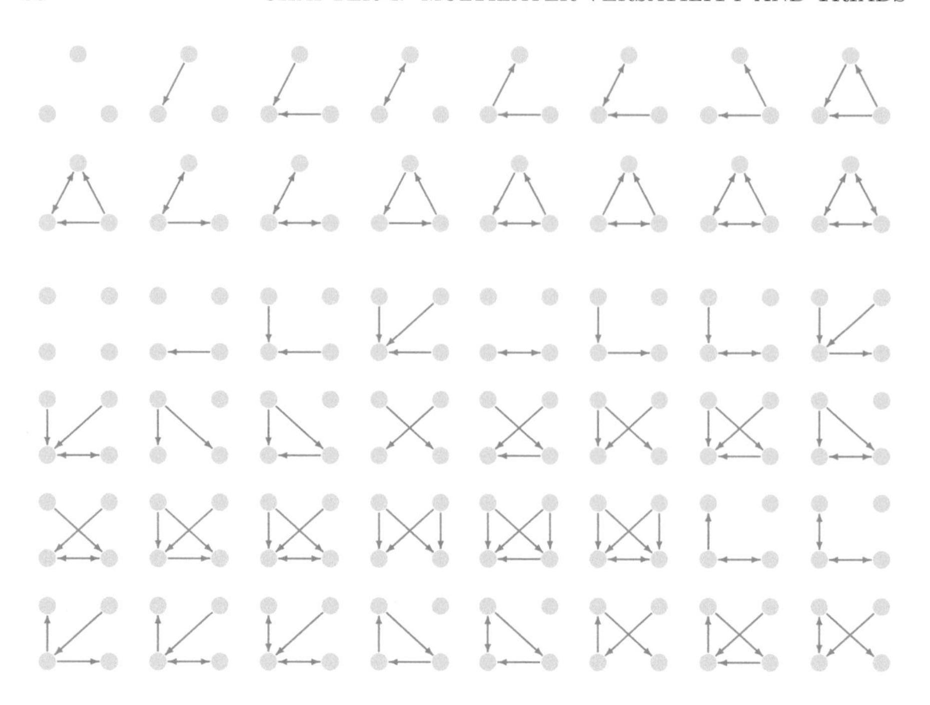

Figure 4.13: The 16 motifs that can be built by directed connections between 3 nodes.

Figure 4.14: Sample of 32 motifs, out of 218, that can be built by directed connections between 4 nodes.

to reveal if specific patterns, such as self-regulation, feed-forward loops, regulating or regulated feedback loops, cascades, bifan, cross-regulation and others characterize information exchange in a complex system.

The analysis of motifs is, nowadays, among the most relevant ones in molecular biology, where they describe small-scale circuits responsible for information processing and response to stimuli within a cell. However, assessing the significance of motifs is computationally expensive, with a computational complexity that rapidly grows with the size of the system and the size of the subnetworks considered. Therefore, it is not surprising that this challenge has been mostly tackled by computer scientists and bioinformaticians, who were among the first ones to attack the problem of finding motifs in edge-colored and node-colored networks representing biological and molecular systems [264].

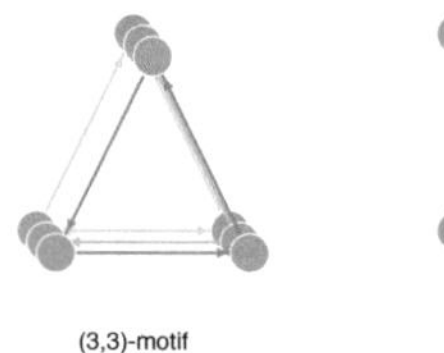

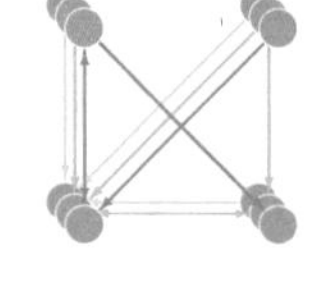

Figure 4.15: Example of multilayer motifs: 3 layers (colored edges), 3 (left) and 4 (right) nodes. This type of motifs is obtained by composing, simultaneously, single-layer motifs. Figure from [4] under Creative Commons Attribution-ShareAlike 4.0 International License.

In the multilayer framework, motifs might consist of directed edges from different layers simultaneously, therefore they are usually indicated as (n, L)–motifs, where n is the number of nodes in the subnetwork and L the number of layers. Figure ***4.15*** shows two representative examples of a $(3, 3)$–motif and a $(4, 3)$–motif.

Up to date, the FANMOD algorithm developed by Sebastian Wernicke and Florian Rasche is the fastest approach to multilayer motif analysis [264] and it is the one used in **muxViz** .

▸ **Code snippet 4.4**
`multi_motifs.R`

4.3 Multilayer triadic closure

Of special interest in applications are those network motifs corresponding to triadic closure, cycles of order 3 where any walk starts and ends in the same node. From this definition, it is clear that not all 3–motifs encode triadic closure (Fig. ***4.13***). The analysis of triadic closure is relevant for applications and it has been used to explain the small-world phenomenon observed in social and biological systems [20], as well as to understand the stability of social interactions evolving over time [265], to mention a few representative examples.

▸ **Code snippet 4.5**
`example_transitivity_new.R`

Triadic closure is extended to multilayer networks by observing that cycles can be closed through multilayer walks and, consequently, triangles can be

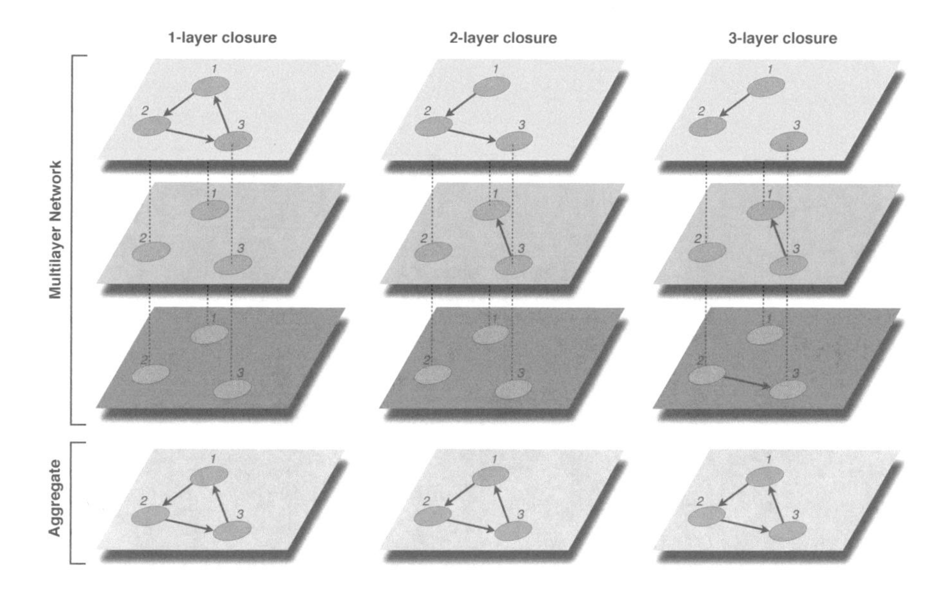

Figure 4.16: Triadic closure in a multilayer network consisting of 3 layers. The aggregate network is not able to distinguish if the closure happens in only one layer or because of the coupling between multiple layers. Figure from [4] under Creative Commons Attribution-ShareAlike 4.0 International License.

closed either in one layer or through the coupling of multiple layers [129, 151, 248] (Fig. ***4.16***).

In fact, the number of triads $n_\Delta(aggr)$ in the aggregate network is mathematically an upper bound to the number of unique triads $n_\Delta(single)$ that one can find from each layer separately and the number $n_\Delta(mux)$ of triads in the multilayer network. An effective method to find multilayer triads is to first find triads in each layer separately and in the aggregate network: triads closed across layers are the ones which are accounted for in the aggregate network but not in the single layers.

The random expectation can be calculated by considering an Erdös-Rényi network with the same number of nodes and edges in each layer and in the aggregate representation. Let $p(\ell) = 2|E|(\ell)/(N(N-1))$ be the corresponding wiring probability[18] for layer ℓ: the random expectation for the number of triads is given by $\binom{N}{3}p^3(\ell)$. The same argument holds for the aggregate network, and the results can be used to quantify the relevance of multilayer triadic closure with respect to single-layer and aggregate triadic closure.

[18] This calculation assumes an undirected network with no multiple edges and self-loops. If this is not the case, the definition should be modified accordingly.

Chapter 5 | Multilayer Organization: Meso-scale

Contents

ONE of the most important goals of network science, and specifically of multilayer network analysis, is the identification of the mesoscale organization of a system in terms of connected components (or clusters) and modules (also known as groups or communities), since it provides a coarse-graining description of the underlying network in terms of structural or functional mesoscopic units.

5.1 Multilayer connected components

The analysis of connected components plays an important role in network science [23, 24], because they identify clusters of nodes who are able to exchange information. More technically, two nodes belongs to the same connected component if there is a path between them. In the case of directed networks different types can be considered:

- **Strongly connected component**: each node is reachable from any other node in the same component or, equivalently, a directed path exists between any pair of nodes, even when origin and destination are swapped;
- **Weakly connected component**: if the undirected representation of the component is considered nodes are part of the same component. In general, if a directed path between a pair of nodes exists, the opposite is not be true for at least one pair.

If the size of the system is finite, the cluster consisting of the largest number of nodes is defined *largest connected component* (LCC), whereas it is named *giant connected component* if the network has infinite size. In general a network can have multiple connected components, which are disjoint clusters of nodes: if there is only one connected component then the network is defined to be *connected*.

Historically, the first class of multilayer networks studied by means of connected components is the one of interdependent networks [141]. In this class of models, two systems A and B are interconnected with links and the potentially functional clusters are identified by *mutually connected components*. If we indicate by $\mathcal{A}$ the set of nodes in network $G(A)$ and by $\mathcal{B}$ the corresponding set of nodes in network $G(B)$, they form a mutually connected component if: i) each pair of nodes in $\mathcal{A}$ is connected by a path consisting of nodes belonging to $\mathcal{A}$ and links of network $G(A)$, and ii) each pair of nodes in $\mathcal{B}$ is connected by a path consisting of nodes belonging to $\mathcal{B}$ and links of network $G(B)$ [139].

In the case of other multilayer networks classes we have seen that a multilayer path consists of a sequence of state nodes across layers. We can use the definition

M. De Domenico, *Multilayer Networks: Analysis and Visualization*, https://doi.org/10.1007/978-3-030-75718-2_5

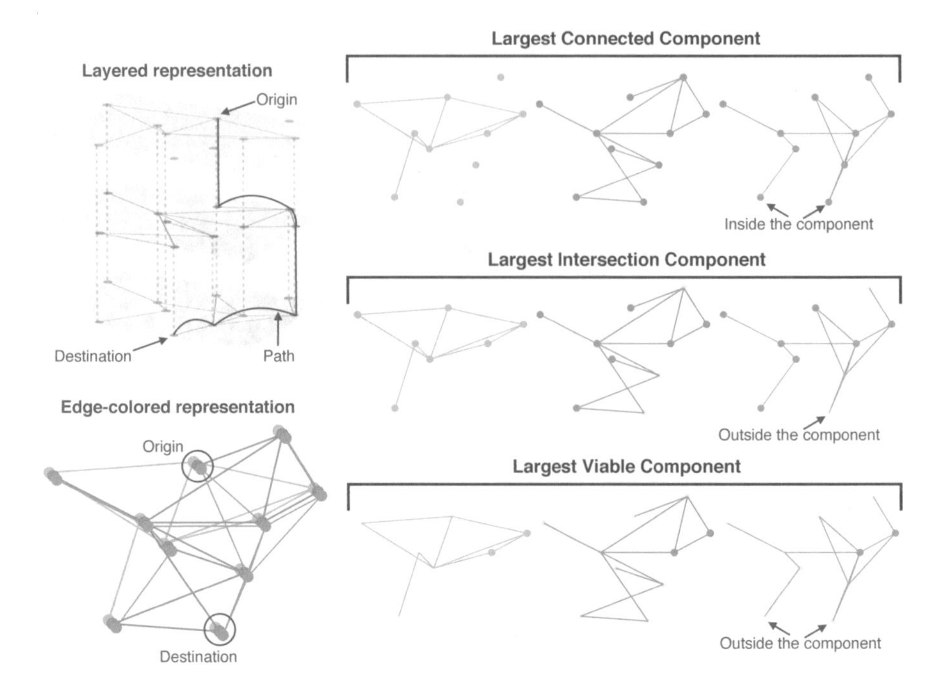

Figure 5.1: Example of a multilayer network represented as an interconnected multiplex (top-left) and an edge-colored multigraph (bottom-left). Multilayer connected components are identified by paths between state nodes (see the text for details). The right-hand side panels highlight the nodes in the largest connected component (top), in the largest intersection component (middle) and the largest viable component (bottom). Figure from [4] under Creative Commons Attribution-ShareAlike 4.0 International License.

given in Sec. 3.5 to define a multilayer connected component as a set of nodes connected by a multilayer path [18]. As for classical networks, it is possible to define strongly and weekly multilayer connected components if connectivity is directed.

The above definition allows to identify the connected components of physical nodes from the aggregate representation of the multilayer network, because it is sufficient that two physical nodes are connected by a path to be part of the same cluster and this property is preserved in the aggregate network.

However, more restrictive definitions are possible. For instance, one might be interested in identifying the largest cluster in which nodes are connected across all layers simultaneously [226]: this cluster is known as the *largest intersection component* (LIC) and can be identified by intersection the LCC of each layer separately. This definition provides a more restrictive selection of nodes than the LCC. Another alternative, recently used to better understand the emergence of continuous or abrupt percolation phase transitions even in systems of finite size, is to aggregate the multilayer system with respect to the intersection of edges and then identify the largest connected component of the resulting network [230].

▸ **Code snippet 5.1**
`connected_components.R`

An even more restrictive selection is achieved by requiring that nodes are *viable*, i.e., they maintain connections in every layer to other viable nodes. In practice, the *largest viable component* (LVC) consists of nodes that are connected by a path in each layer simultaneously [226]. As a result, all nodes in the LVC are essential to the function of the system and to define its structural core [257, 266, 267]. The more restrictive condition imposed by the LVC is responsible for a hybrid phase transition which leads to the discontinuous emergence of the giant viable cluster, at variance with ordinary percolation where a continuous phase transition is observed [226].

Clearly, the size of the LVC is equal or smaller than the size of the LIC, which in turn is equal or smaller than the size of LCC (FIg. ***5.1***), all of them providing different and complementary information about the structure of a multilayer network.

Another emblematic example of the the versatility offered by multilayer modeling is given by the integration of social and ecological data to analyze the behavior of socio-ecological systems. It has been recently proposed to use layers

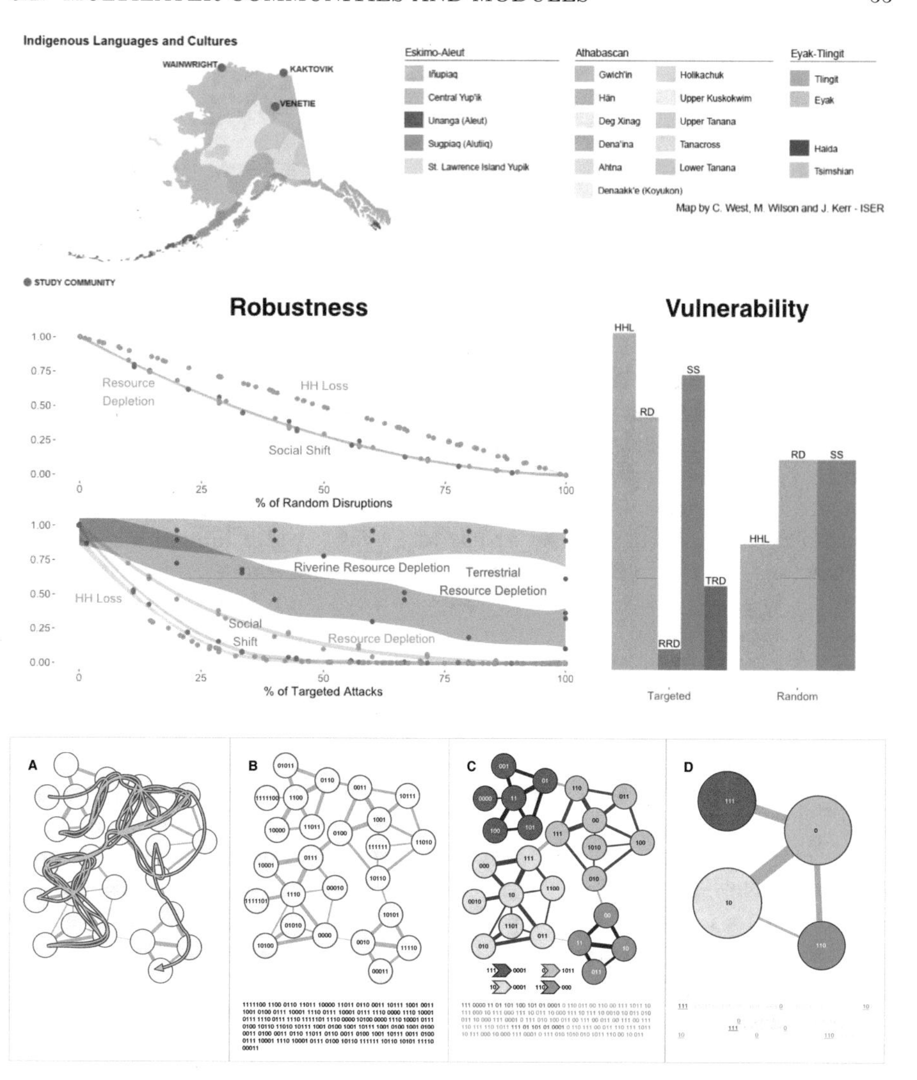

Figure 5.2: Analysis of the vulnerability of socio-ecological networks to plausible scenarios of change. **Top left**: study areas in northern Alaska. **Bottom Left**: Robustness of the multilayer network of the three communities (band) to random and targeted perturbations. **Bottom right**: Expected vulnerability of the three communities to the targeted and random removal due to social shift (SS), resource depletion (RD), terrestrial resource depletion (TRD) and riverine resource depletion (RRD), as well as key households (HHL). Reproduced from Ref. [8]

Figure 5.3: Community detection on classical networks using the Infomap algorithm. (a) The trajectory of a random walk on a network is shown with a solid line. (b) The trajectory can be efficiently described by encoding each node with a codeword – a string defined by a sequence of zeros and ones (bits), e.g., 01011 – of the Huffman codebook shown in the panel. For instance, if the walker starts from the node in the upper left corner, the corresponding string will start with 1111100, and if the second visited node is the neighbor on the right-hand side, 1100 is added. The whole trajectory in (a) is described by the 314 bits shown in the bottom of the panel. (c) When the partition encoded by colors is used to describe the random walk dynamics, the trajectory encoded by the corresponding codebook is described, on average, by 32% less information, because the walk within each group in this partition is more persistent over time and less bits are required for its description. (d) This knowledge can be used to effectively and efficiently coarse-graining the network into functional modules. Image reproduced with permission from Ref. [13], Copyright (2008) National Academy of Sciences, U.S.A.

to encode the unique combinations of ecological resources and social relations, with the aim of representing community networks [8]. A direct application to three small geographically isolated communities in Alaska, revealed surprising results. The study involved two coastal Iñupiat communities, Wainwright (553 people) and Kaktovik (239 people), whose subsistence depends on bowhead, beluga, caribou, and other marine and terrestrial species, and one interior Athabascan Gwich?in community, Venetie (166 people), whose subsistence centers on moose, caribou, salmon, and other riverine and terrestrial species. In fact, those communities are exposed to several disturbances caused by extreme climate change and industrial development. Despite their vulnerability to resource depletion, multilayer analysis (Fig. *5.2*) surprisingly revealed that robustness is mostly undermined by the loss of key households and the erosion of cultural ties linked to sharing and cooperative social relations [8].

5.2 Multilayer communities and modules

One of the most important problems in network analysis is the identification of groups of nodes with a special role for the structure or the function of a system. In fact, many empirical systems represented by classical networks are well organized into *clusters* (or *modules* or *groups* or *communities*, all terms that are used in different disciplines to refer to the same concept). It has been extensively

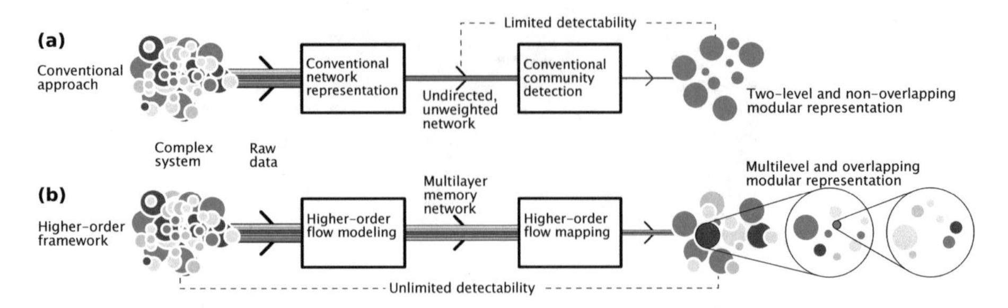

Figure 5.4: A schematic illustration of how data from complex systems is processed using (a) conventional approaches – often reducing data into unweighted and undirected networks – and (b) higher-order approaches based on flow modeling. Besides being more accurate, the second class of methods allows for the detection of richer functional information than standard approaches. Image reproduced with permission from Ref. [14].

shown that social and biological networks are characterized by this *mesoscale organization* with important structural and functional meaning [71, 84] quantified by their modularity. Historically, the first methods for community detection have been developed in social science and were based on the assumption that groups and their inter-group connectivity can be modeled by stochastic blocks [268–270], an hypothesis extensively adopted also in computer science for machine learning [271–274]. Among the other methods proposed by network scientists, there are the ones based on exploiting how information flows and is trapped within and between modules [85, 88] or how its compression from an information-theoretic perspective can be used to identify functional modules [13, 86, 275]. Even a brief description of all other methods available in the literature would be well beyond the scope of this book, therefore we refer the interested reader to the thorough reviews by the outstanding network scientists who played, with others, a central role for the development of this field [55, 60, 276].

In this section, instead, we describe multilayer community detection. Several methods have been proposed to cope with the additional level of complexity introduced, for instance, by multiplexity of units and their interactions. Chronologically, one of the pioneering work on this topic has been published by Mucha et al [87], who introduced multi-slice modularity maximization – generalizing the modularity function [74] widely used for community detection in classical networks – for applications to interconnected multiplex and temporal networks, later improved [277, 278]. A different method, based on tensor factorization, has been introduced for the analysis of time-varying systems and applied in the context of contact networks [279]. Generative models rooted on statistical physics and Bayesian inference have been introduced more recently [238, 280] to generalize another traditional approach – widely used in social science and computer science – based on stochastic block models [268, 269]. Studies have shown the enhanced detectability of community structure in multilayer networks when layers are aggregated [281], for instance, by summation, highlighting the possibility to uncover otherwise hidden communities by combining layer aggregation with thresholding techniques [282].

Here, we focus our attention on another method, based on the analysis of random walk dynamics on the top of the network from the perspective of information theory, named `Infomap` [13]. A schematic illustration of how `Infomap` works[19] is given in Fig. ***5.3***. Its generalization, known as `MultiplexInfomap`[20] and distributed with `muxViz`, has been recently introduced [15] and better understood within the framework of higher-order flow modeling [14, 283–285] (Fig. ***5.4***).

As its classical counterpart, `MultiplexInfomap` is based on the compression of network flow modeled through random walk dynamics: the flow is compressed when regular structures – i.e., functional multilayer modules – are present, and the procedure is mathematically described by the *map equation*[21]. The generalization is natural, because the same information-theoretic machinery is applied to the non-Markovian flow characterizing higher-order models with memory, with the memory of the present layer playing a role similar to memory due to previous steps. In this framework, a multilayer community iden-

[19] An interactive app is available at the URL http://www.mapequation.org/apps/MapDemo.html#applet

[20] An interactive app is available at the URL http://www.mapequation.org/apps/sparse-memory-network/index.html

[21] http://www.mapequation.org/

tifies a group of nodes persistently capturing network flows within and across layers (Fig. ***5.5***).

In practice, a random walk dynamics with specific properties is defined: jumping between nodes within the same layer is Markovian, whereas switching across layers is non-Markovian. The dynamics of random walks in this context is described, with some detail, in Box 5.2.1.Figure ***5.6*** shows the typical result of a mesoscale analysis[22].

[22] Note that the version of InfoMap assumed in the LIB is the latest one available and it is different from the one included with the GUI, which is based on an older version.

Box 5.2.1: Transition tensors used in `MultiplexInfomap`

In the case of multilayer networks where layers are structurally coupled, the entries of the rank–4 transition tensor governing the random walk are defined

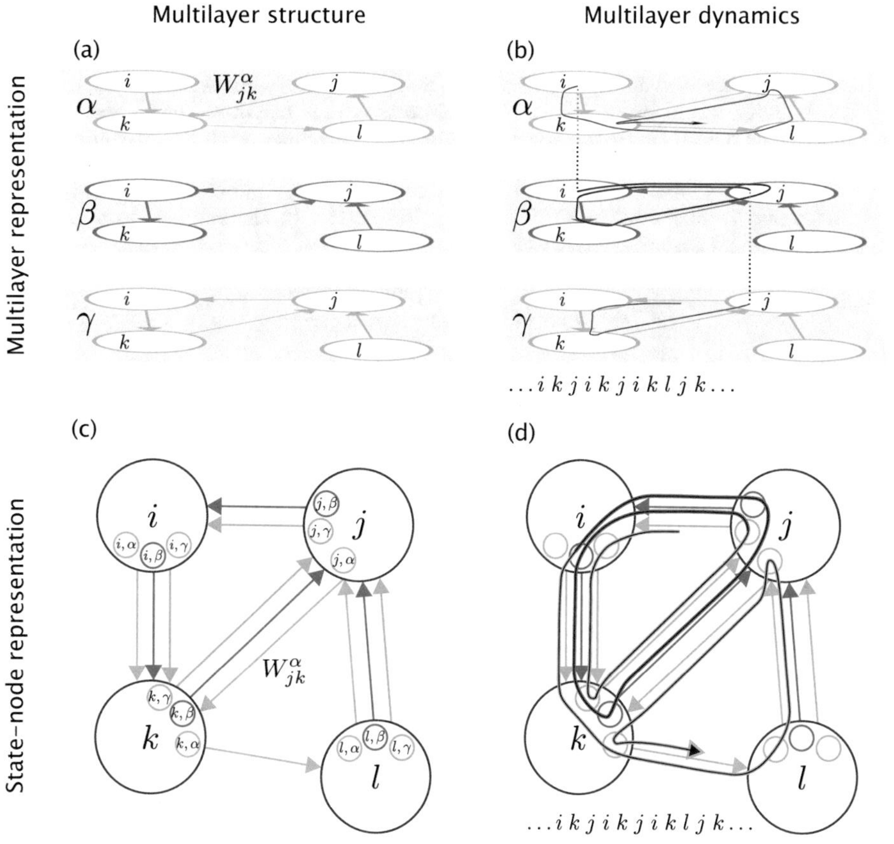

Figure 5.5: Higher-order flow dynamics on a multilayer network consisting of three layers and four physical nodes. (a) Structural representation of the system. (b) The trajectory of a random walker on the multilayer network, jumping and switching between state nodes, while its dynamics through physical nodes is recorded. (c) The system represented in terms of the network induced by its state nodes (encoded by colors); state nodes exist within the corresponding physical nodes (encoded in black). (d) The same trajectory shown in (c) but in the state-node representation, where only the sequence of visited physical nodes is recorded. Image reproduced with permission from Ref. [15].

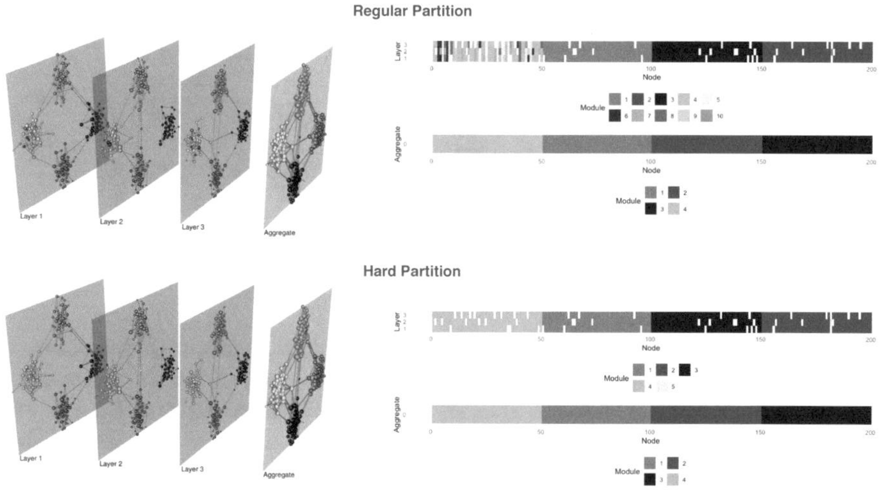

Figure 5.6: Multilayer community detection with regular (top panels) and hard (bottom panels) partitions for the whole multilayer network and its aggregate representation.

by

$$\mathcal{P}_{ij}^{\alpha\beta} = \frac{D_{(i)}^{\alpha\beta}}{S_i^{(\alpha)}} \frac{W_{ij}^{(\beta)}}{s_i^{(\beta)}}. \quad (5.1)$$

Let us remark here that the above equation is not tensorial: this definition is chosen for its readability and easy interpretability. In fact, $W_{ij}^{(\beta)}$ indicates the intra-layer adjacency matrix of layer β and $s_i^{(\beta)} = \sum_{j=1}^{N} W_{ij}^{(\beta)}$ is the out-strength of node i in that layer, while $D_{(i)}^{\alpha\beta}$ indicates the adjacency matrix of inter-layer links between any pair of layers from the perspective of node i and $S_i^{(\alpha)} = \sum_{\beta=1}^{L} D_{(i)}^{\alpha\beta}$ is the corresponding inter-layer out-strength [18]. Therefore, the probability of a transition is given by the product of two independent transition probabilities: i) the one of changing layer from α to β and ii) the one to jump from node i to node j once in layer β.
However, when information about inter-layer connectivity is missing or the system is represented by a non-interconnected multiplex network, the switching probability would be undefined. For this reason it is useful to model the dynamics across layers by means of a random walker which, with a probability r named *relax rate*, is forced to change layer by switching through replica nodes. Consequently, with probability $1-r$ the random walker jumps between nodes belonging to the same layer. Therefore, the entries of the rank–4 transition tensor in this case are defined by

$$\mathcal{P}_{ij}^{\alpha\beta}(r) = (1-r)\,\delta_{\alpha\beta}\frac{W_{ij}^{(\beta)}}{s_i^{(\beta)}} + r\frac{W_{ij}^{(\beta)}}{S_i}, \quad (5.2)$$

with $S_i = \sum_{\beta=1}^{L} s_i^{(\beta)}$ the overall out-strength of node i and $\delta_{\alpha\beta}$ the Kronecker delta.
Note that Eq. (5.1) and Eq. (5.2) are equivalent if $D_{(i)}^{\alpha\beta} = (1-r)\,\delta_{\alpha\beta}S_i + rs_i^{(\beta)}$ and $S_i^{(\alpha)} = \sum_{\beta=1}^{L} s_i^{(\beta)}$. We refer the interested reader to Ref. [15] for further details about the properties of relaxed dynamics.
It is worth remarking that the relax rate r is a parameter: changes in the mesoscale organization, such as community splitting and merging, are expected for varying r. An information-theoretic method to identify the value of the relax rate for which the corresponding mesoscale structure is the most representative has been recently proposed [16]. The method is based on an independent cost function known as *normalized information loss*, corresponding to the log-likelihood of a stochastic block model. For a specific relax rate r, this cost function is defined by

$$H_r(X|Y) = \log_2\left[\prod_{i=1}^{m}\prod_{j=1}^{m}\binom{n_i n_j}{l_{ij}}\binom{w_{ij}-1}{l_{ij}-1}\right], \quad (5.3)$$

being n_i the number of nodes in cluster i, l_{ij} and w_{ij} indicating, respectively, the number of links and the total weight of links between clusters i and j. The normalized version of the cost function is used for practical purposes:

$$H_r^*(X|Y) = \frac{H_r(X|Y) - \min\limits_{0<r\leq 1} H_r(X|Y)}{\max\limits_{0<r\leq 1} H_r(X|Y) - \min\limits_{0<r\leq 1} H_r(X|Y)}. \quad (5.4)$$

It can be shown that, by minimizing the normalized information loss, it is possible to learn the latent block structure in presence of real-valued weights through the use of a parametric distribution, as shown in [286], and the optimal compression of the system is achieved.

▸ **Code snippet 5.2**
`community_detection.R`

To better understand how to use **`MultiplexInfomap`** with the normalized information loss (NIL) to select the most representative partition of the system, we consider four distinct duplex networks (Fig. *5.7*) and show how the number of identified communities and NIL change with the relax rate (Fig. *5.7*). Results show that there exist a type of critical threshold r_{thr} above which information

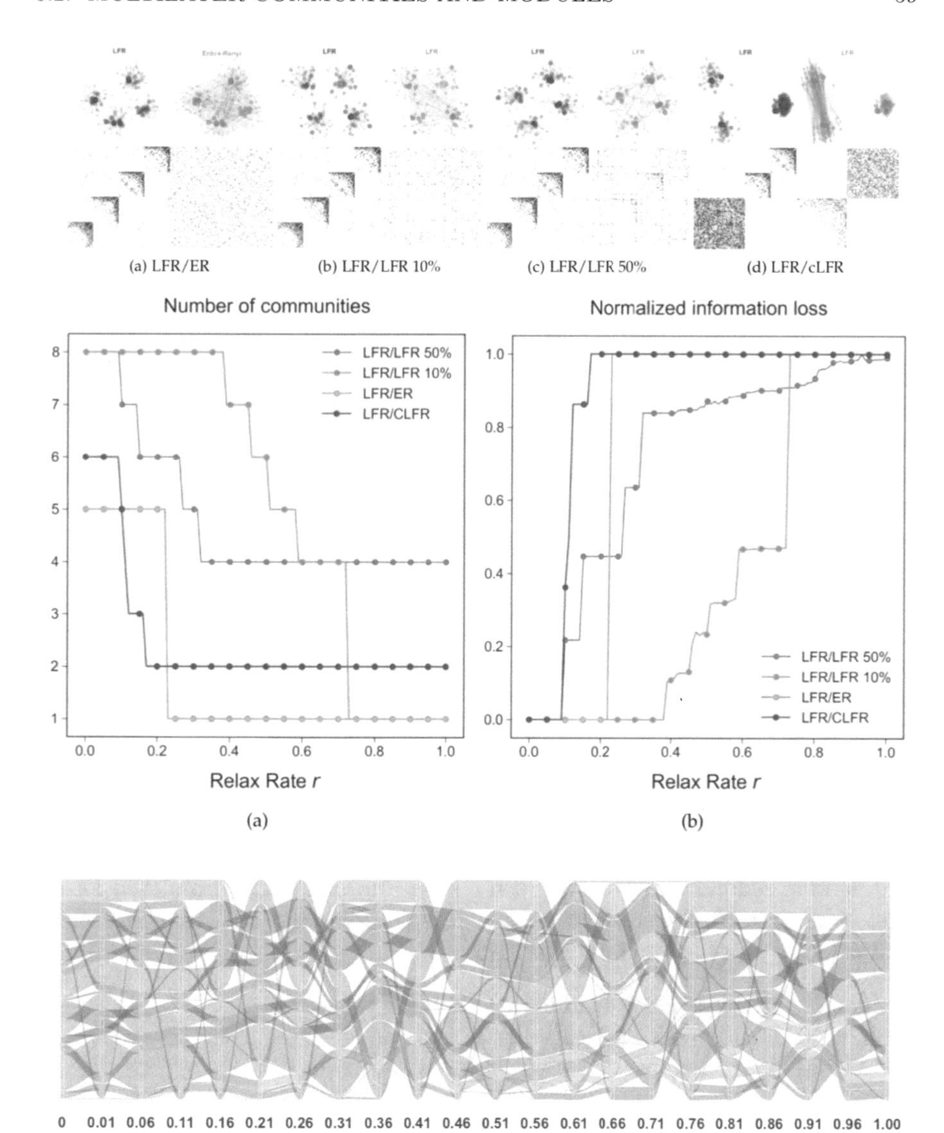

Figure 5.7: **Top:** Two layers of distinct duplex networks, represented as graphs (top panels) and adjacency matrices (bottom panels). 128 nodes with a community structure (Lancichinetti-Fortunato-Radicchi model) in layer 1 are coupled to (a) topological noise (Erdős-Rényi model) or a LFR network with 10% (b) and 50% (c) of nodes belonging to overlapping communities. In (d) the complementary network (cLFR) is considered. Figure from [16]. **Bottom:** Number of communities (a) and normalized information loss (b) *versus* the relax rate r for the same multiplex networks. Figure from [16]

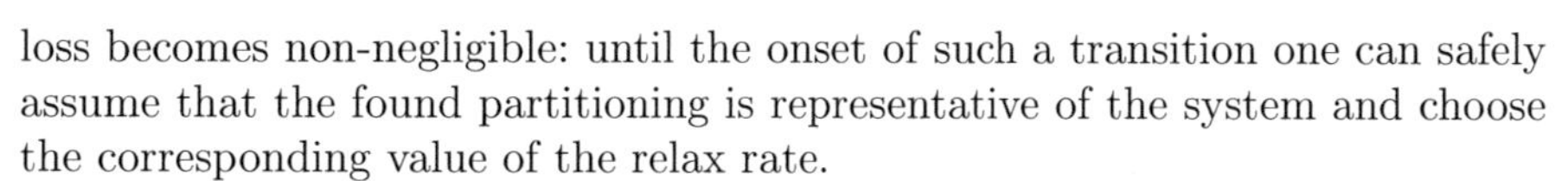

Figure 5.8: Community structure (obtained from `MultiplexInfomap`) of the human multiplex proteome for increasing values of relax rate r. Only modules with more than 100 proteins are shown: the visualization provides insights about how modules split and merge *versus* r. Figure from [16].

loss becomes non-negligible: until the onset of such a transition one can safely assume that the found partitioning is representative of the system and choose the corresponding value of the relax rate.

As a direct application of these ideas to empirical systems, let us consider the human interactome where nodes are proteins, links encode their interactions and layers represent different types of interactions (such as physical, chemical and genetic). When `MultiplexInfomap` is applied, one has to choose a specific value of the relax rate or, alternatively, scan over different values of this parameter to see how the resulting mesoscale organization changes accordingly.

The result of the latter approach is shown in Fig. **5.8**, where it is possible to get a visual understanding about the reorganization of the largest groups for increasing relax rate. Fig. **5.9** shows the distribution of clusters size for distinct values of the relax rate, to gain insights about how smaller groups merge into larger ones for increasing r.

Remarkably, when one considers the partition obtained for the optimal value of the relax given by the NIL it has been shown that the corresponding functional content – from a biological perspective – is the highest. This analysis, based on standard enrichment through the Molecular Signatures Database[1]

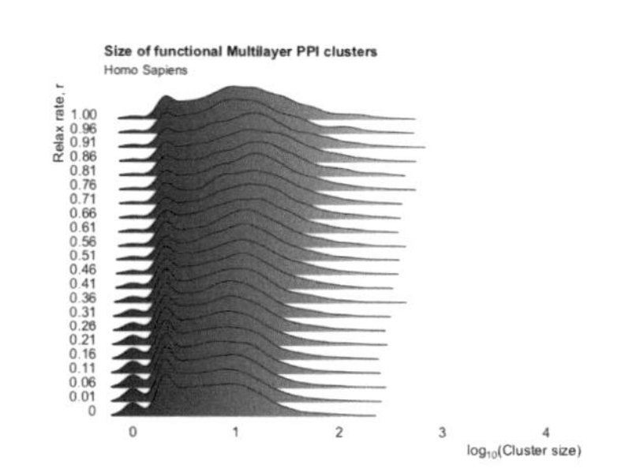

Figure 5.9: Same analysis of Fig. **5.8**, to show the distribution of clusters? size *versus* r. When the relax rate increases the number of very small clusters decreases. Figure from [16].

[1] http://software.broadinstitute.org/gsea/msigdb/collections.jsp

(MSigDB), highlights how this method is suitable to gain insights about complex multilayer biological systems.

5.3 Clustering and reducing multilayer structures

Multilayer networks are powerful tools for modeling empirical complex systems and integrating different sources of information although, in general, it should be common practice to ask whether or not the multilayer representation is the most suitable representation for the network under investigation. In fact, there is mathematical evidence that the layers of interconnected multiplex systems, for instance, under certain conditions can be analyzed in isolation [193]. The analysis of the structural and dynamical properties of many empirical systems suggests that disentangling relationships in social networks [248, 287–289], ecological systems [9, 154–156], transportation systems [18, 290], molecular networks [16, 291–294] and human brain [146, 295–300], is often desirable.

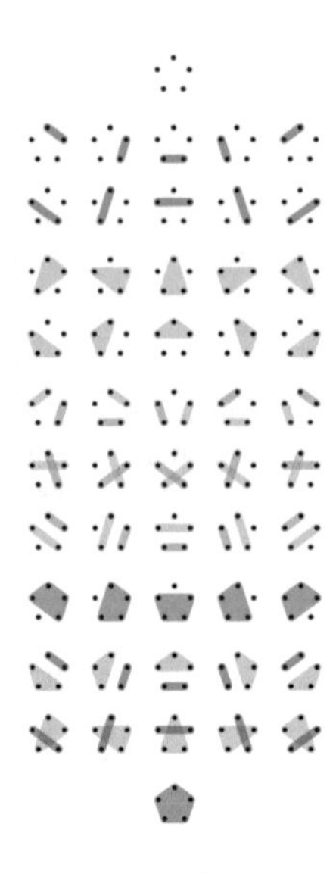

Figure 5.10: Partitions of 5 elements into 52 different groups. Reproduced from Wikipedia under Creative Commons Attribution 3.0 Unported License https://creativecommons.org/licenses/by/3.0/

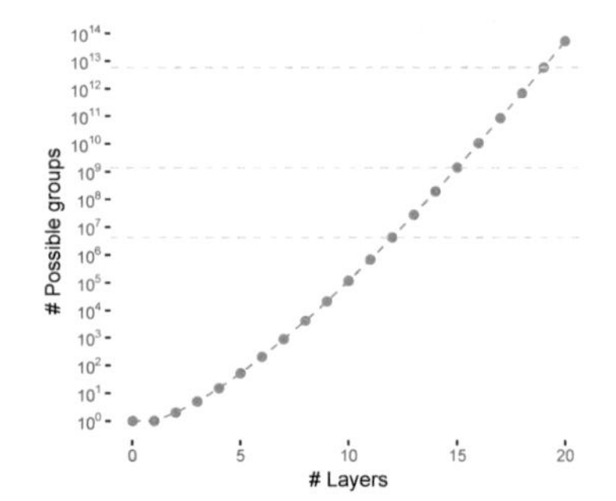

Figure 5.11: The number of partitions scales super-exponentially and it is known as the Bell number. Exhaustive searches on multilayer networks consisting of more than 10 layers are computationally prohibitive.

Therefore, for modeling and analytical purposes it is useful to devise a strategy to understand to which extent the multilayer representation is needed or if single-layer or aggregate representations are valuable alternatives. For instance, consider the extremal case of a multiplex system with identical layers: it would be desirable to aggregate *structurally redundant layers* according to a reduction strategy. In fact, redundancy, quantified in terms of overlapping edges, is very common [150, 153]. We wonder if it is possible to account for more complex redundant topological patterns, not necessarily the one based on edge overlapping, and if a procedure preserving relevant topological information while aggregating redundancies exists. Such a procedure has been formulated for the first time in 2015, it is known as *structural reducibility* [301] (see Fig. ***5.12***) and consists of a few fundamental steps:

1. Identify layers which are similar according to specific criteria;
2. Devise a strategy to aggregate layers together:
3. Control the procedure by means of a cost function which penalizes the aggregation of structurally different layers and favors the aggregation of layers characterized by topological redundancy.

More recently, a procedure for functional reducibility has been proposed: formally, it is based on the same steps described above while variations apply to the criteria for aggregating layers and calculating entropies (see Fig. ***6.2***).

Step 1. requires to define a (dis)similarity measure between any pair of layers. This task can be achieved by using a distance measure: if two layers are distant they are also dissimilar, otherwise they are similar and good candidates to be aggregated without loss of information. In the original approach [301], the quantum Jensen-Shannon divergence has been used (see Box 5.3.1 and Box 5.3.2 for details).

Box 5.3.1: Information entropy of a complex network

Defining and calculating the information entropy of a complex network is, in general, a difficult task. Many approaches are based on extracting specific descriptors – e.g., degree – and their (possibly joint) probability distribution and calculating its Shannon entropy. However, this type of approaches only considers a few descriptors, providing a limited knowledge of the underlying network and, consequently, a biased estimation of its information entropy.
An alternative, proposed in 2009 [302], is based on encoding network's structural information in a new operator $\rho^i_j = L^i_j/2E$ – named density matrix, because it shares the same properties of the well known operator in Quantum Mechanics, where L^i_j is the combinatorial Laplacian matrix of the network and $2E$ encodes the total number of edges – and calculate the Von Neumann entropy as

$$S(\rho) = -\mathrm{Tr}(\rho^i_j \log_2 \rho^j_k) = -\sum_{i=1}^{N} \lambda_i \log_2 \lambda_i \text{ bits}, \tag{5.5}$$

where λ_i is the i–th eigenvalue of the density matrix and Tr is the trace operator. This can be generalized to the case of interconnected multilayer systems [129]. More recently, it has been shown that a more suitable definition for the density matrix is given by

$$\rho^i_j = \frac{e^{-\tau L^i_j}}{Z_\tau}, \qquad Z_\tau = \mathrm{Tr}(e^{-\tau L^i_j}) = \sum_{i=1}^{N} e^{-\lambda_i \tau}, \tag{5.6}$$

which encodes information entropy in terms of a network discovery process based on diffusion [303], the parameter τ playing the same role of a Markov time.
It has been suggested [301] to define the entropy S_{mux} of a multilayer systems in terms of the entropies S_ℓ ($\ell = 1, 2, ..., L$) of each layer in isolation as

$$S_{\mathrm{mux}} = \frac{1}{L} \sum_{\ell=1}^{L} S_\alpha(\rho_\alpha), \tag{5.7}$$

corresponding to the average entropy of layers. More recently, it has been show that this entropy is not suitable to describe the information content of multilayer systems characterized by strong correlations among layers and a more adequate definition must be used instead [17] .

Box 5.3.2: Distance between complex networks

Von Neumann entropy and the formalism inspired by quantum computing can be used to define the quantum Jensen-Shannon divergence between two networks consisting of the same number of nodes. This quantity is the generalization of the well known Jensen-Shannon divergence widely adopted in information theory, because it is symmetric and allows for the definition of a metric distance, at variance with the Kullback-Leibler divergence. Given two networks – for instance, the ones corresponding to two distinct layers of a multilayer system – with density matrices ρ^i_j and σ^i_j, respectively, their network Jensen-Shannon divergence [301] is quantified by

$$\mathcal{D}_{JS}(\rho||\sigma) = S\left(\frac{\rho + \sigma}{2}\right) - \frac{S(\rho) + S(\sigma)}{2} \text{ bits}, \tag{5.8}$$

whereas the Jensen-Shannon distance is given by

$$d_{JS}(\rho, \sigma) = \sqrt{\mathcal{D}_{JS}(\rho||\sigma)}, \tag{5.9}$$

which is minimum (i.e., $d_{JS} = 0$) when the knowledge of one network is sufficient to gain maximum information about the other one – or, equivalently, the two networks are perfectly identical – and maximum ($d_{JS} = 1$) otherwise. Remarkably, this metric can be used to compare all pairs of layers in a multilayer system to build a distance map that can be used, for instance, to cluster layers hierarchically [301, 303] into communities. The same metric has been also used to compare an empirical system against multiple network models with the purpose of fitting the corresponding parameters and perform model selection [303].

Step 2. is far from being trivial. In fact, from an algorithmic perspective, one should compare all possible aggregations of layers in groups of any size, estimate the redundancy and decide which combination of layers preserves most of the original information. However, considering all possible aggregations requires is computationally expensive, since the possible number of groups out of L layers is given by the Bell number, which scales super-exponentially with L (see Fig. ***5.10*** and ***5.11***). In fact, a good greedy strategy is to map layer-layer similarities into pairwise distances and use the resulting distance matrix to perform a hierarchical clustering of layers. Closest layers in the hierarchy are the most likely candidates for the aggregation, resulting in at most $L-1$ possible

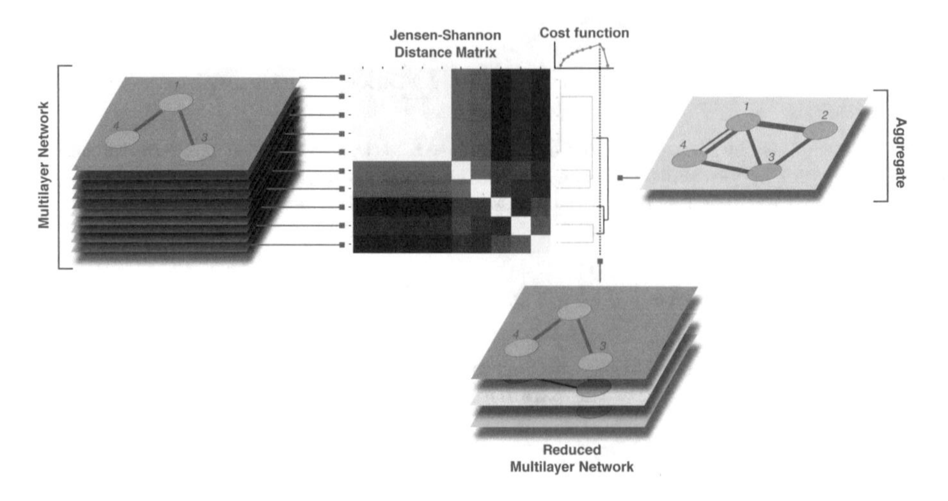

Figure 5.12: Schematic illustration of structural reducibility analysis of a multilayer network consisting of $L = 10$ layers with redundant information (here encoded by repeated colors). Layers are hierarchically clustered according to their Jensen-Shannon distance. The clustering is used for consecutive aggregations of closest layers and the cost function increases because, in this example, no information is loss at each step. The process continues until the optimal configuration is reached, i.e., when only $L' = 3$ non-redundant layers are found. Figure from [4] under Creative Commons Attribution-ShareAlike 4.0 International License.

calculations. The aggregation by sum is usually used, but other methods are possible.

Step 3. is required to understand where to stop aggregating. This is achieved by introducing the cost function

$$q(m) = 1 - \frac{S_{\text{mux}}(m)}{S_{\text{agg}}}, \tag{5.10}$$

where $m = 0, 1, 2, ..., L-1$ is an index encoding the number of aggregated layers ($m = 0$ corresponds to the whole multilayer network), $S_{\text{mux}}(m)$ is the entropy of the system at step m and S_{agg} is the entropy of the fully aggregate network. Since $S_{\text{mux}}(m) \leq S_{\text{agg}}$ for any m, the cost function is bounded within $[0, 1]$, and it can be shown that it reaches a maximum value if and only if the multilayer network is distinguishable from its full aggregation, i.e., the system is characterized by structurally redundant layers [301]. If the network can be fully aggregated, the cost function is expected to monotonically decrease for increasing m. In fact, aggregations that alter the information content of the underlying topology are disfavored, whereas this is not the case when topologically redundant layers are aggregated.

▸ **Code snippet 5.3**
`structural_reducibility.R`

Chapter 6 | Other Multilayer Analyses Based on Dynamical Processes

Contents

In this chapter we briefly mention some dynamical processes defined on the top of multilayer networks, which are relevant to get insights about how easy it is to navigate a complex structure and how to obtain functionally reduced representation of a multilayer system. While the topics briefly discussed in this chapter are not exhaustive, and many others will be added in future editions of this book, they are available in `muxViz` or can be easily implemented.

6.1 Navigability of multilayer systems

In practical applications, from efficient routing of information to fast exploration of a neighborhood, it is often useful to quantify how much a network is navigable [304, 305]. To this aim, it is possible to exploit random walks to define the *coverage* $\rho(t)$ of a system, i.e., the average fraction of nodes visited at least once within a certain time t. In the case of multilayer networks, a physical node is considered as visited if among its state nodes at least one has been visited by walkers [18].

The evolution of the coverage provides useful information about the navigability of the system at different temporal scales. In fact, let $p_{i\alpha}(t)$ be the vector[1] encoding the probability to find a random walker in a node i of layer α and let $p_i(t) = p_{i\alpha}(t)u^{\alpha}$ indicate the probability to find a random walker in node i at time t regardless of its origin and of the layer where the observation is taken. The master equation is clearly

$$p_{i\alpha}(t+1) = p_{i\alpha}(t)\mathcal{P}^{i\alpha}_{j\beta}, \tag{6.1}$$

▸ Code snippet 6.1
`example_coverage.R`

The scalar probability to find the walker in node a at time $t+1$, regardless is given by the projection $p(t+1; a) = p_{i\alpha}(t+1)e^{i}(a)u^{\alpha}$.

By assuming that the random walker originated in node b, let $\sigma(t; b \to a)$ indicate the probability to *not* find the walker in node a after t time steps. It is possible to show [18] that the equation

[1] Formally this is not a vector but can be flattened to a supra-vector and, with the abuse of notation discussed in the previous chapters, the (i, α)–th entry of this supra-vector corresponds to the probability of finding the walker in node i of layer α.

M. De Domenico, *Multilayer Networks: Analysis and Visualization*, https://doi.org/10.1007/978-3-030-75718-2_6

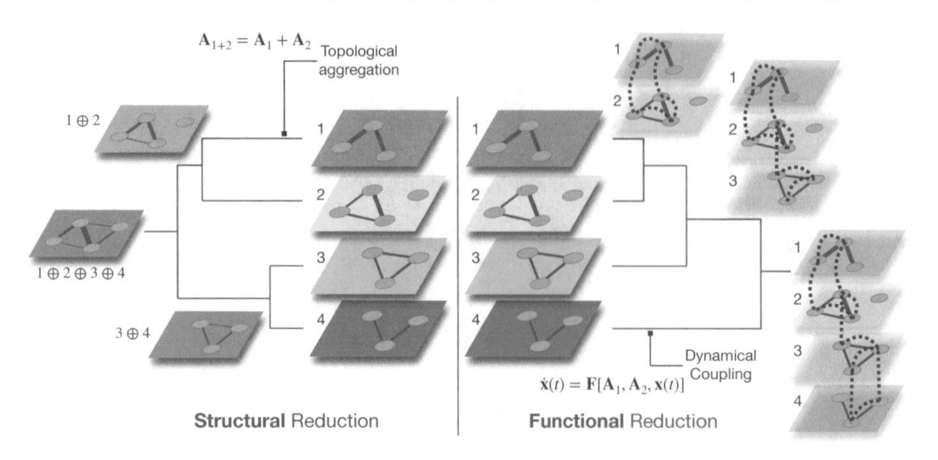

Figure 6.2: Schematic illustration comparing structural against functional reduction of a multilayer network consisting of $L = 4$ layers. The procedure is similar for both approaches, but the relevant difference is that while structural reducibility alters the topology of the system, the function reducibility allows to functionally coupling layers without altering their structure. Figure from [17].

$$\sigma(t+1; b \to a) = \sigma(t; b \to a)\,[1 - p(t+1; a)] \tag{6.2}$$

is satisfied, provided that $\sigma(0; b \to a) = 1 - \delta(a, b)$, where $\delta(a, b)$ is the Kronecker function. Therefore $\sigma(0; b \to a) = 1$ if origin and destination do not coincide, and it is zero otherwise, since a walk starting at node b cannot be at the same time at node a unless $b = a$. The solution of the above recursive equation is given by

$$\sigma(t; b \to a) = \sigma(0; b \to a) \exp\left[-e_i(b)e_\alpha(1)\mathbb{P}^{i\alpha}_{j\beta}(t)e^j(a)u^\beta\right], \quad \mathbb{P}^{i\alpha}_{j\beta}(t) = \sum_{\tau=0}^{t} (\mathcal{P}^{\tau+1})^{i\alpha}_{j\beta}, \tag{6.3}$$

where, without loss of generality, the factors $e_i(b)e_\alpha(1)$ account for the assumption that the walker started in node b and in the first layer[2] at time $t = 0$. The tensor $\mathbb{P}^{i\alpha}_{j\beta}(t)$ encodes the probability to reach each node through any path of length $1, 2, \ldots, t+1$.

The evolution of network coverage can be approximated with remarkable accuracy by double averaging over all nodes the probability $1 - \sigma(t; b \to a)$:

$$\rho(t) = 1 - \frac{1}{N^2} \sum_{\substack{a,b=1 \\ a \neq b}}^{N} \exp\left[-e_i(b)e_\alpha(1)\mathbb{P}^{i\alpha}_{j\beta}(t)e^j(a)u^\beta\right]. \tag{6.4}$$

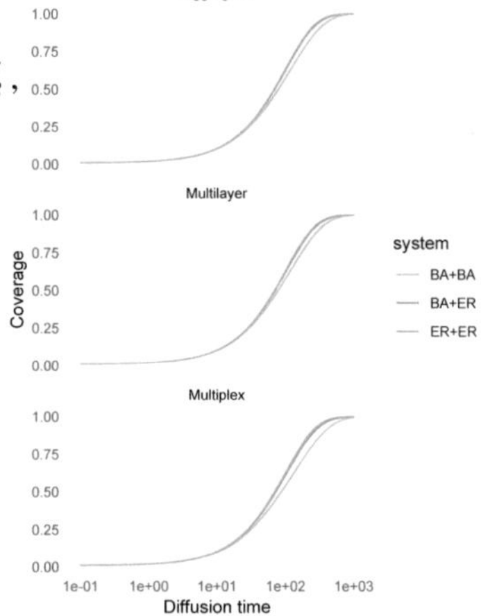

Figure 6.1: Evolution of the coverage of a multilayer system consisting of 3 layers and 100 nodes. Different layers are characterized by different combination of topologies: Barabasi-Albert (BA) and Erdös-Rényi (ER). Different models are used: from top to bottom, respectively, the coverage of the aggregate, the interconnected multiplex and the non-interconnected multiplex networks is shown. Genuine multilayer effects emerge from the interplay between structure and dynamics: the navigability of the overall system depend on how different layers are coupled together [17, 18].

The coverage is sensitive to network topology and transition rules used to navigate the system. In fact, nodes are visited at different time scales: direct effects include a slower or faster exploration of the system with respect to its layers separately or its aggregate network (Fig. ***6.1***).

6.2 Functional reducibility of multilayer systems

In the case of functional reducibility, the coupling between layers is considered with respect to dynamics, specifically random walks [17] because they provide a powerful proxy for a broad spectrum of transport properties. This method is not yet available in **muxViz**, therefore we limit to mention its existence as a reference for future development and refer to Fig. ***6.2*** to illustrate the difference between functional and structural reducibility methods.

[2] Remind that $e_i(b) \in \mathbb{R}^N$ is the b–th canonical vector in the space of nodes and $e_\alpha(r)$ is the r–th canonical vector in the space of layers.

Chapter 7 Visualizing Multilayer Networks and Data

Contents

A fundamental step for the analysis of multilayer networks is visual inspection of the underlying structure, as well as to encode multivariate information extracted from a system into a simple-yet-effective visual representation. In this chapter we will briefly describe how to embed nodes and layers in a three-dimensional space, to visualize a multilayer network in terms of floating layers, as well as to show the results of multidimensional analysis by using the annular visualization.

7.1 Embedding nodes and layers in a 3D space

The visualization of multilayer networks is not a trivial task. At variance with single-layer systems, there is more information to account for and to show, in principle the same amount for each layer separately. One possibility, when the

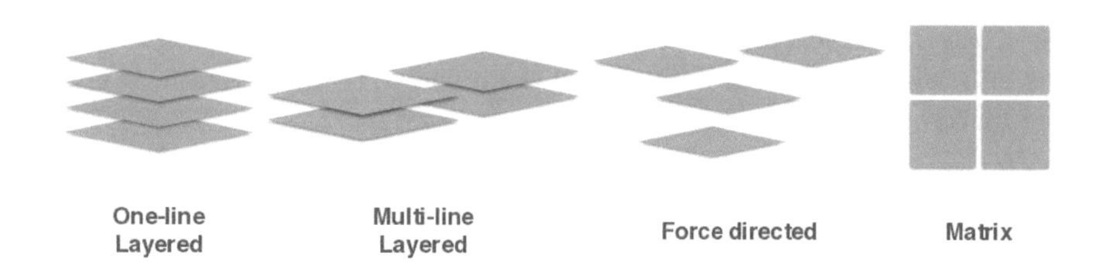

Figure 7.1: Different arrangement of layers in a three-dimensional space. Each square encodes a distinct layer and multiple layers are organized to highlight some underlying multilayer pattern. The most adopted structure is the "one-line layered". Figure from [4] under Creative Commons Attribution-ShareAlike 4.0 International License.

M. De Domenico, *Multilayer Networks: Analysis and Visualization*, https://doi.org/10.1007/978-3-030-75718-2_7

number of layers is small, is to show each layer independently from the others, as we have already done in some cases in this book: the obvious drawback is that in this way it is difficult to identify patterns *across* the layers, requiring more cognitive efforts. Another possibility is to plot a heatmap of the corresponding supra-adjacency representation: also in this case the cognitive effort increases for increasing network size and number of layers.

A more convenient procedure to visualize multilayer systems is to show their layered structure in 2.5 dimensions, something in between 2D and 3D. The advantage of this approach is that when layers are aligned according to specific criteria (e.g., to favor the alignment of replica nodes across layers) the cognitive effort to understand the multilayer patterns is dramatically reduced. Of course there are multiple ways to embed layers in a three-dimensional space, as shown schematically in Fig. *7.1*. Both the GUI and the LIB version of `muxViz` provide simple ways to create this type of visualization, the "one-line layered" structure being the most widely adopted nowadays.

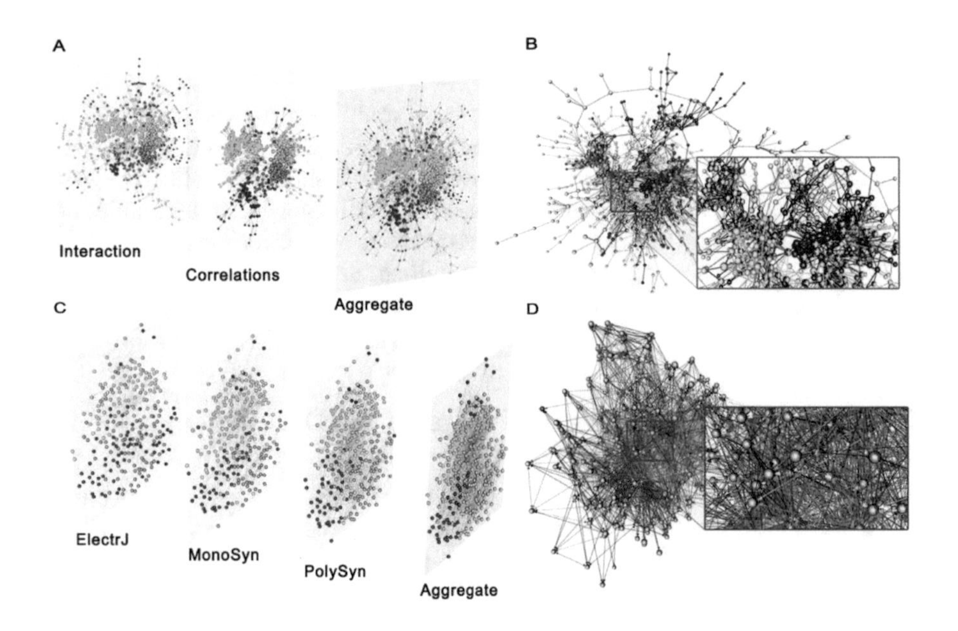

Figure 7.2: Two distinct methods – (A-C) one-line layered and (B-D) 3D edge-colored – to represent a multilayer network (in this case an edge-colored multigraph), for two empirical systems: (A-B) interaction and correlation-based network of genes in S. cerevisiae, the common yeast, and (C-D) different synaptic junctions – electric (ElectrJ), chemical monadic (MonoSyn) and polyadic (PolySyn) – characterizing the connectome of C. elegans. See the text for details. Figure from [19].

This approach works quite well if the position of nodes is calculated while taking advantage of the whole multilayer information. There are some ways to achieve this goal, the easiest one being based on applying standard force-directed layout algorithms – such as Kamada-Kawai [306], Fruchterman-Reingold [307] or distributed recursive layout (DRL) [308] – on some aggregate representation of the network, e.g., the one obtained by the union or the intersection of edges across layers.

A valid alternative, when the number of nodes is not too large, is to encode distinct interactions by the color of links and, accordingly, to color nodes with the colors of the layers where they are connected: in this case, interesting visualizations can be obtained by applying the force-directed layout algorithms in three dimension instead of two. We show a representative example in Fig. *7.2*, where multilayer genetic and neuronal networks are considered. On the left-hand side the one-line layered representation is used, with node colors encoding their organization into groups or functional modules. On the right-hand side the edge-colored three-dimensional representation is shown, where colors encode distinct type of interactions. Clearly, both types of representation have advantages and disadvantages: the topological patterns visibile in the latter do not allow to disentangle the contribution of each layer, like in the former, and by using colors to encode the interaction type one has to sacrifice other types of information that could be shown (e.g., the community membership).

Note also that in case of inter-layer connectivity, the layered method allows one to explicitly visualize links, whereas this is not possible in the edge-colored representation.

Other methods, like the one based on using diffusion geometry to embed a network into a diffusion space [89], are currently under investigation.

7.2 Annular visualization of multivariate data

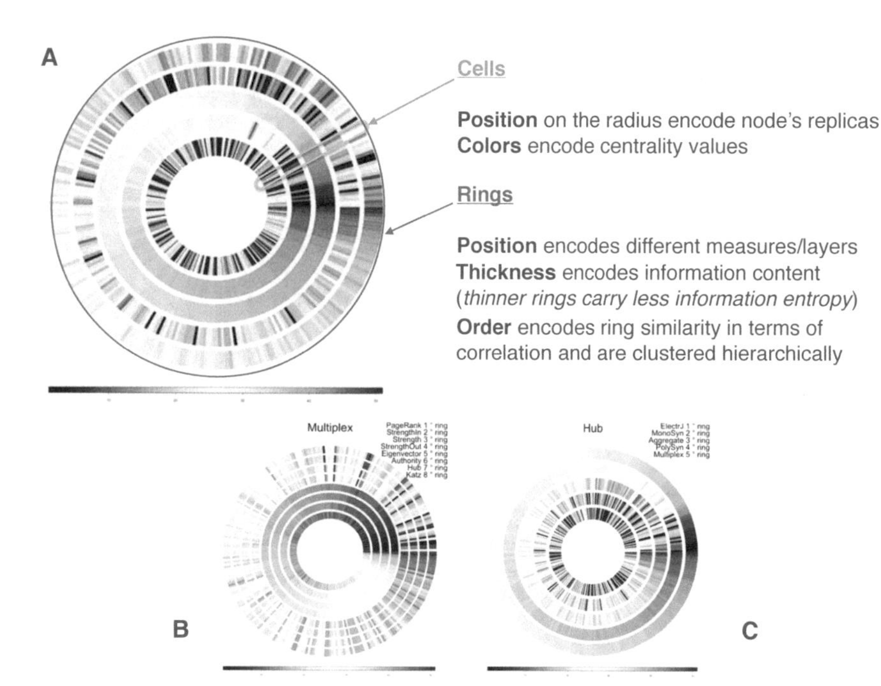

Figure 7.3: Annular visualization for the visual analysis of multivariate information produced by multilayer analysis. See the text for further details.

Once the multilayer analysis of a complex network is performed, it can be challenging to represent and visualize the wealth of information obtained. A way to visualize the multivariate information resulting from this type of analysis is provided by the *annular visualization*, which is designed to allow one to easily capture patterns and deduce qualitative information about the multilayer descriptors produced. More specifically, the annular visualization in `muxViz` GUI focuses on the comparison of centrality and versatility descriptors obtained from the analysis of single layers, the aggregate network and the multilayer network, respectively, although the underlying approach can be easily generalized to visualize and compare other descriptors.

Figure 7.3A shows a typical annular visualization, consisting of concentric rings – each one encoding a specific vector of information, such as a centrality or a versatility profile – where each cell is uniquely identified by an angle and encodes a unique node. The color of a cell is representative of the value of the descriptor corresponding to a specific node.

The position of rings can encode different types of information. The first type is to calculate several multilayer versatility descriptors, encode each one into a ring and compare them (Fig. 7.3B). The second type is to focus on one descriptor, for instance HITS, and calculate the corresponding centrality profile in each layer separately, in the aggregate network and in the multilayer network, defining the rings for comparison (Fig. 7.3C). Both approaches are important to gain insights about the correlation among different versatility profiles (first type) or the difference in the role(s) played by nodes when layers are considered in isolation, aggregated, or analyzed through the multilayer lens (second type). The annular visualization is a fundamental tool to gain a broad spectrum of visual insights including, for instance, to understand if the versatility of a node is primarily due to its centrality in a specific layer or if a single layer of the aggregate network might be considered as a good proxy for the whole multilayer structure.

The thickness of a ring is proportional to the information content it encodes: if the distribution of values in a ring is strongly peaked around a certain one, the corresponding information entropy is low and one does not learn much by visualizing that ring with full size. To reduce visual noise the thickness of that ring is reduced accordingly, whereas it is maximized when information it carries is high.

The order of rings is also important and can be customized to maximize the readability of the visualization. For instance, in `muxViz` 's default setting, the distance between each pair of descriptors is calculated by means of a measure of correlation (e.g., Pearson, Spearman, or JS divergence) and the result is used to hierarchically cluster the corresponding rings. This information is used to arrange less distant vectors into adjacent rings.

Correction to: Multilayer Networks: Analysis and Visualization

Manlio De Domenico

Correction to:
M. De Domenico, *Multilayer Networks: Analysis and Visualization*, https://doi.org/10.1007/978-3-030-75718-2

The book was inadvertently published with an incorrect prior affiliation for the author, Manlio De Domenico, as "Universitat Rovira i Virgili, Tarragona, Spain". It has been updated as "University of Padua, Italy" in the front matter.

The updated online version of this book can be found at
https://doi.org/10.1007/978-3-030-75718-2

M. De Domenico, *Multilayer Networks: Analysis and Visualization*, https://doi.org/10.1007/978-3-030-75718-2_8

Part II
Appendices

Chapter A | Installing and Using muxViz

A.1 The "R Universe" of muxViz

muxViz comes with two distinct versions:

- v2.0 allows one to use a GUI and depends on some packages which, in turn, depends on many other packages which create the R Universe of the software (Fig ***A.1***). The advantage of this version is that it does not really requires one to know how to code in R, but it does not allow for low-level control of functions but it facilitates the analysis and visualization of multilayer networks.
- v3.1 is a standalone library (LIB) which does not allow to use a GUI. Therefore it is much easier to install than v2.0 and easy to control from one's own scripts: however, it requires the knowledge of R to be used.

The LIB version is strongly recommended, since it is faster to install and allows one for a broader set of analytical and computational tools. Moreover, it is the only version that will be maintained in the future.

A.2 Requirements and Installation for v3.1 (LIB)

[23] http://www.r-project.org/

As mentioned in the previous section, muxViz requires at least R[23] v3.2.0 (or above) to be installed and work properly. However, since some dependencies might work only with more recent versions of R, also muxViz, in principle, will require a recent R version. This is true at least for the GUI version (2.0), whereas the standalone LIB version (3.1) builds on less packages and, consequently, should be compatible with a broader range of versions.

Since the analysis of complex multilayer networks is mostly based on operations with matrices and vectors, it is strongly recommended to:

[24] http://www.netlib.org/lapack/

[25] http://www.netlib.org/blas/

[26] https://mran.microsoft.com/download/

- either install the R environment and configure the system to support efficient linear algebra libraries, such as LAPACK[24] and BLAS[25];
- or install the enhanced R environment developed by Microsoft named Microsoft R Open[26].

Provided that the devtools package is installed, the following step is not required:

```
1 install.packages(devtools)
```

[27] https://github.com/manlius/muxViz/

muxViz LIB is available as an open source framework and its source code can be download for free from Github[27] or, within the R environment, directly installed by

```
1 devtools::install_github("manlius/muxViz")
```


M. De Domenico, *Multilayer Networks: Analysis and Visualization*, https://doi.org/10.1007/978-3-030-75718-2

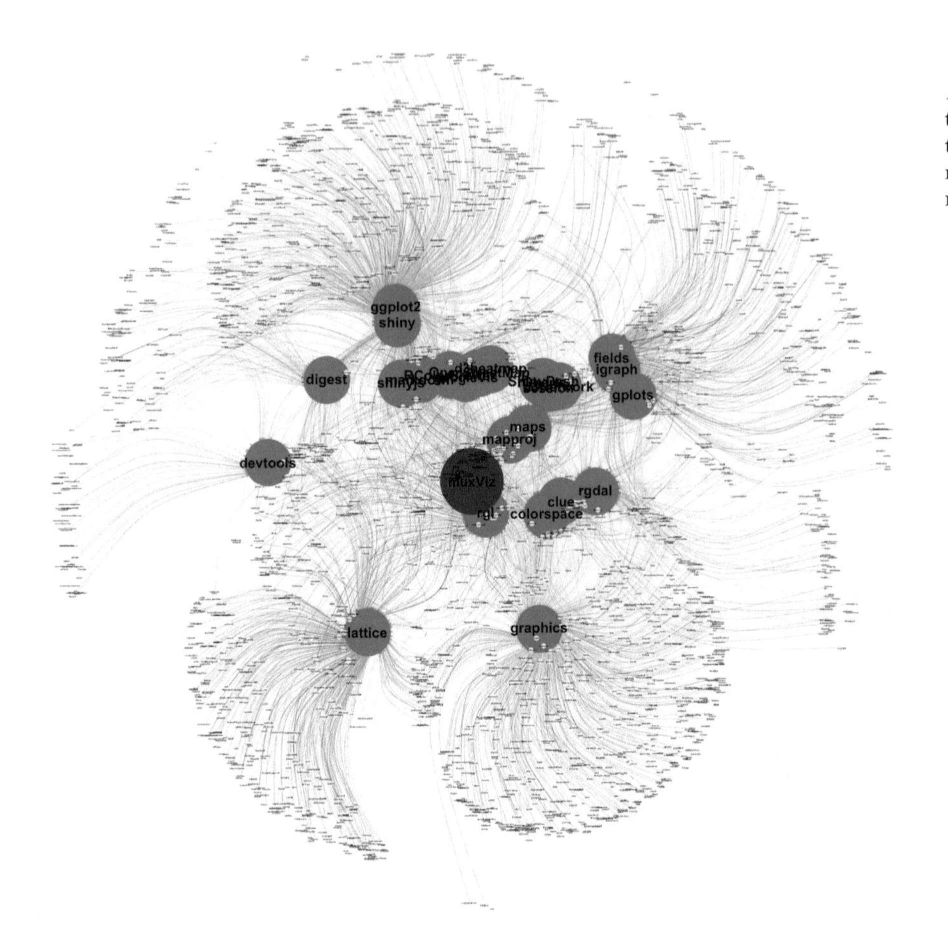

Figure A.1: Network representation of dependencies of `muxViz` 2.0: theblue markers indicate main direct depedencies, while smaller markers indicate indirect ones.

In the same repository, the full package[1] documentation – with usage tips and examples – is provided, together with some vignettes on multilayer community detection, multilayer network motifs and on how to setup the Infomap and FANMOD routines. Additional scripts with examples are provided.

A.3 Requirements and Installation for v2.0 (GUI)

`muxViz` requires at least R[28] v3.2.0 (or above) to be installed and work properly. However, since some dependencies might work only with more recent versions of R, also muxViz, in principle, will require a recent R version. This is true at least for the GUI version (2.0), whereas the standalone LIB version (3.1) builds on less packages and, consequently, should be compatible with a broader range of versions.

[28] http://www.r-project.org/

Since the analysis of complex multilayer networks is mostly based on operations with matrices and vectors, it is strongly recommended to:

- either install the R environment and configure the system to support efficient linear algebra libraries, such as LAPACK[29] and BLAS[30];
- or install the enhanced R environment developed by Microsoft named Microsoft R Open[31].

[29] http://www.netlib.org/lapack/

[30] http://www.netlib.org/blas/

[31] https: //mran.microsoft.com/download/

Once the R environment is installed, some additional libraries have to be installed to allow for the installation of the required R packages. Users should verify that their systems have a working installation of Java[32] and GDAL[33]

[32] https://www.java.com/

[33] http://www.gdal.org/

[1] We acknowledge the precious help of Dr. Giulia Bertagnolli to setup and build this package.

(Geospatial Data Abstraction Library), required for visualization of geographical networks. Java is required to allow R the automatic installation of `rJava` and `OpenStreetMap` packages, whereas GDAL is required to install `sp` and `rgdal` R packages. More information about possible issues are reported in troubleshooting file accompanying **`muxViz`** .

`muxViz` is available as an open source framework and its source code can be download for free from Github[34]. **`muxViz`** is able to detect the required R packages which are still missing and install them automatically: therefore, it is unlikely that the user will need further action.

[34] https://github.com/manlius/muxViz/

Once **`muxViz`** has been downloaded, it can be unzipped[35] anywhere in user's system, e.g. in `/user/path/muxviz`. Open the R environment and set the working directory to **`muxViz`** by typing:

[35] Note that *git* can be used instead.

```
1 setwd("/user/path/muxviz")
```

and then simply type:

```
1 source("muxVizGUI.R")
```

to start the framework. While **`muxViz`** is loading, messages useful for debugging anomalous scenarios are printed in the R terminal. Note that only the first time **`muxViz`** is loaded, it will try to install all the missing R packages it requires for functioning correctly: therefore, on older systems, the first load might require a few minutes.

If everything has been installed correctly, **`muxViz`** will open a new page on the default Web browser where a splash screen like the one shown in Fig. ***A.2*** should appear:

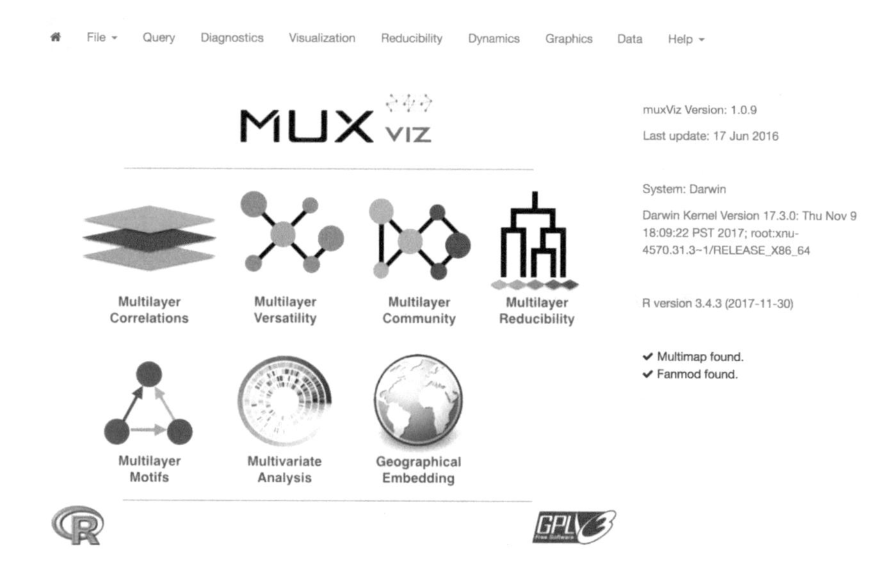

Figure A.2: Splash screen of **`muxViz`** as rendered by any standard Web browser. A visual summary of available modules is shown in the middle, as well as information about the system (right panel).

A.4 Troubleshooting

`muxViz` can count, nowadays, on an online community with more than 600 members. This section is dedicated to some problems and solutions identified with the help of this community.

A.4.1 Very quick installation on GNU/Linux

If you use a Linux (Ubuntu-like) distribution, you are very lucky, because the following BASH script will do the job for you:

```
#download R from their repository
wget http://cran.es.r-project.org/src/base/R-3/R-3.0.3.tar.gz
DIR=$PWD

#install R
sudo apt-get build-dep r-base-core
sudo mv R-3.2.0.tar.gz ~
cd ~
tar xvf R-3.2.0.tar.gz
cd R-3.2.0
./configure
make
sudo make install

#install GDAL
sudo apt-get install libgdal1-dev libproj-dev
```

A.4.2 Ubuntu 14.04

One user reported the following solution for installation on this system. To load muxViz 2.0 in R environment, he solved by modifying the muxVizGUI.R file:

Before:

```
1 devtools::install_github("shiny-incubator", "rstudio")
```

After:

```
1 devtools::install_github("rstudio/shiny-incubator", "rstudio")
```

A.4.3 Multimap or FANMOD not found

muxViz uses an API to external softwares: Multimap (known as multiplex infomap) and FANMOD. The home screen of muxViz will indicate if the available binaries are correctly installed and can be used by the platform. If it is not the case, a WARNING message is generally displayed.

Should one care about missing FANMOD and/or Multimap? It depends. If Multimap is missing, you will not be able to perform multilayer community detection based on the generalization of the Infomap algorithm for this type of networks. If FANMOD is missing, you will not be able to perform multiplex motif analysis of your network.

If one is interested in using these features, and the default installation shows a WARNING, one just needs to compile the software on their own machine. This is an easy task on all OSs.

- Verify that you have a c++ compiler on your machine. On GNU/Linux and most recent versions of Mac OS X, it is already installed. In Windows, you might need some extra work (http://www.mingw.org/wiki/howto_install_the_mingw_gcc_compiler_suite, https://www.youtube.com/watch?v=k3w0igwp-FM). You might want to take a look at the official web site to installing GCC (https://gcc.gnu.org/wiki/InstallingGCC, https://gcc.gnu.org/install/specific.html)

- Go into the 'src' folder that is placed inside the **muxViz** folder
- Unzip the software ('fanmod_src.zip' and/or 'Multiplex-Infomap_src.zip')
- Enter into the software folder from the terminal and run 'make'. This step is expected to go smoothly, because after a few seconds the GCC compiler will produce binary executables *ad hoc* for your system. If it is not the case, take a look at the official web page http://www.gnu.org/software/make/manual/make.html and if you still can not compile, post on our Google Group
- Copy the produced binaries inside the 'bin' folder that is placed inside the **muxViz** folder
- Verify that the software is correctly named as 'fanmod_linux' and/or 'multiplex-infomap_linux'

A similar procedure, except for the last point, is required to use the same tools with the standalone library .

A.4.4 Possible errors when using Motifs

The issue is that there is a conflict between the current version of igraph and shinyjs. More details can be found at https://github.com/igraph/igraph/issues/846. While waiting for a new release of igraph, solving the issue, we can install the dev version:

```
1 devtools::install_github("igraph/rigraph")
```

On Mac OS X this could be a bit tricky, because R might use clan instead of gcc/g++ to compile. A solution is to create the file

`~/.R/Makevars`

if it does not exist, and set the following parameters:

```
CFLAGS +=               -O3 -Wall -pipe -pedantic -std=gnu99
CXXFLAGS +=             -O3 -Wall -pipe -Wno-unused -pedantic

VER=-4.2
CC=gcc$(VER)
CXX=g++$(VER)
SHLIB_CXXLD=g++$(VER)
FC=gfortran
F77=gfortran
MAKE=make -j8
```

Restart R and try again to install the dev version of igraph.

A.4.5 Possible errors with rgdal

To work properly with geographical networks, the GDAL (Geospatial Data Abstraction Library) is required and should be installed before running **muxViz** for the first time. GDAL should be available as an R package and should be easily installed just by typing

```
1 install.packages("sp")
2 install.packages("rgdal")
```

within the R environment. However, in a few cases it can be more complicated and some users reported problems for its installation. If this is also your case you might want to take a look at some suggestions on Stackoverflow (http://stackoverflow.com/questions/15248815/rgdal-package-installation) or on spatial.ly (http://spatial.ly/2010/11/installing-rgdal-on-mac-os-x/).

In any case, it is highly recommend to visit the GDAL website and follow the hints provided there (http://trac.osgeo.org/gdal/wiki/BuildHints).

A.4.6 Possible errors with rjava (any OS)

Some users reported that, when using `muxViz` for the first time, they get the following error:

```
Warning: Error in : package or namespace load failed
for ?OpenStreetMap?:
 .onLoad failed in loadNamespace() for 'rJava', details:
```

One possible solution is to open the terminal and type

```
R CMD javareconf
```

to reconfigure java to work correctly within R. You might read about possible solutions for GNU/Linux (https://stackoverflow.com/questions/3311940/r-rjava-package-install-failing) and Mac OS X (https://github.com/MTFA/CohortEx/wiki/Run-rJava-with-RStudio-under-OSX-10.10,-10.11-(El-Capitan)-or-10.12-(Sierra)).

A.4.7 Possible errors with rjava (latest MacOSs)

It can happen that newest MacOS generate installation issues with rjava. MacOS Users should take a look at the following approaches: http://www.owsiak.org/r-java-rjava-and-macos-adventures/, https://stackoverflow.com/questions/30738974/rjava-load-error-in-rstudio-r-after-upgrading-to-osx-yosemite, http://osxdaily.com/2017/06/29/how-install-java-macos-high-sierra/, https://github.com/MTFA/CohortEx/wiki/Run-rJava-with-RStudio-under-OSX-10.10,-10.11-(El-Capitan)-or-10.12-(Sierra).

In general try make sure R is configured with full Java support (including JDK). Run

```
sudo R CMD javareconf
```

to add Java support to R. If you still can't install it, read below. A possible solution for MacOS versions *before* Sierra:

```
sudo ln -f -s
$(/usr/libexec/java_home)/jre/lib/server/libjvm.dylib
/usr/local/lib
```

But it may happen that your version of MacOS + Java does not have that path. To find the correct path try

```
/usr/libexec/java_home
```

to obtain something like

```
/Library/Java/JavaVirtualMachines/1.6.0.jdk/Contents/Home
```

Look for the dylib file:

```
find $(/usr/libexec/java_home) -name "libjvm"
```

If the result is null, then you might need to install (more) Java. Understand which Java you have installed already and where:

```
/usr/libexec/java_home -V
```

If you see something like this:

```
1.8.0_162, x86_64:  "Java SE 8" /Library/Java/Java...
1.6.0_65-b14-468, x86_64:   "Java SE 6" /Library/Ja...
1.6.0_65-b14-468, i386: "Java SE 6" /Library/Java/J...

/Library/Java/JavaVirtualMachines/jdk1.8.0_162.jdk/...
```

skip the next two paragraphs, otherwise install Java SE 6 and 8.

For Java SE 6 go to Apple support, download: https://support.apple.com/downloads/DL1572/en_US/javaforosx.dmg and install.

Oracle website is the not the best user-friendly website around. Check instructions at https://docs.oracle.com/javase/8/docs/technotes/guides/install/mac_jdk.html and download Java SE 8 from http://www.oracle.com/technetwork/java/javase/downloads/java-archive-javase8-2177648.html.

You should download and install the file

```
jdk-8u162-macosx-x64.dmg
```

or similar, for Java SE 8. Note that it will ask you to register an account: download is free, but you can't skip the registration phase (2 min required).

If Java SE 6 and SE 8 are installed, run again

```
/usr/libexec/java_home -V
```

and hope to see

```
1.8.0_162, x86_64:  "Java SE 8" /Library/Java/Java...
1.6.0_65-b14-468, x86_64:   "Java SE 6" /Library/Ja...
1.6.0_65-b14-468, i386: "Java SE 6" /Library/Java/J...

/Library/Java/JavaVirtualMachines/jdk1.8.0_162.jdk/...
```

so that everything is installed correctly. Then run

```
java -version
```

and hope to see something like

```
java version "1.8.0_162"
Java(TM) SE Runtime Environment (build 1.8.0_162-b12)
Java HotSpot(TM) 64-Bit Server VM (build 25.162-b12, mixed mode)
```

Time to tell R how to use our Java:

```
sudo R CMD javareconf
```

and pray to see at the end of the terminal something like

```
JAVA_HOME         : /Library/Java/JavaVirtualMach...
Java library path: $(JAVA_HOME)/lib/server
JNI cpp flags     : -I$(JAVA_HOME)/../include -I$(JA..
JNI linker flags : -L$(JAVA_HOME)/lib/server -ljvm
Updating Java configuration in /Library/Frameworks..
Done.
```

Now, in the terminal, paste the following:

```
unset JAVA_HOME
R --quiet -e 'install.packages("rJava",
  type="source", repos="http://cran.us.r-project.org")'
```

to install rJava from source with Java 8 JDK. If it works without errors, let's check everything is fine:

```
R --quiet -e 'library("rJava"); .jinit();
.jcall("java/lang/System",
"S", "getProperty", "java.runtime.version")'
```

and you should get something like

```
1.8.0_162-b12
```

as a result.

A.4.8 Install muxViz *with R 3.3 or higher*

muxViz depends on several R packages that, when updated by their developers, might cause issues on **muxViz** . In general, it might happen that you use a very new version of R (**muxViz** was developed for R 3.2): you should still be able to use **muxViz** by simply patching the initial sanity check it does to ensure full compatibility. Open muxVizGUI.R and edit:

```
- Line 1: if(grep("3.3",version$version.string)!=1){
- Line 12: comment it out as
#devtools::install_github("trestletech/ShinyDash")
```

to avoid the error that package ShinyDash is not available (for R version 3.3.0). Finally, install Shiny Dash from R console:

```
1 devtools::install_github("ShinyDash", "trestletech")
```

Now **muxViz** should start and work.

A.4.9 Use existing Linear Algebra Library

On Mac and GNU/Linux it is possible to exploit the already existing linear algebra R packages by forcing R to use a faster BLAS version. On a Mac OS X this is easily achieved by

```
sudo ln -sf /System/Library/Frameworks/
        Accelerate.framework/Frameworks/
        vecLib.framework/Versions/Current/
        libBLAS.dylib  /Library/Frameworks/
        R.framework/Resources/lib/libRblas.dylib
```

A.5 Preparing the data: allowed formats

The data is generally expected to be saved in plain text files known as edge lists, where in general **muxViz** (both the GUI and the LIB versions) expects to find the information about links starting in one node of a specific layer and ending in a node of another (or the same) layer. In the following sections we explain in greater detail the different formats tailored to encode the connectivity information of specific multilayer models.

For the GUI version, the following subsections are more relevant than the LIB version, since in the latter more flexibility for the manipulation of the data structure is allowed. Nevertheless, when using the LIB's routines to import data in **muxViz** format, it is required to follow the standards specified in the following.

A.5.1 Edge-colored networks

The configuration file is a ASCII file including the list of layers to be included in a multiplex, the corresponding labels and the possible layout file to define node properties (e.g., ID, labels, geographic coordinates, etc). The expected format of a configuration file is:

`path_layer_X;label_layer_X;layout_layer_X`

where:

Variable	Type	Mandatory	Description
`path_layer_X`	String	Yes	Specify the path and the filename to the edges list to be used as layer
`label_layer_X`	String	No	Specify the label to be used in the rendering for that layer
`layout_layer_X`	String	No	Specify the path and the filename to the file containing information about nodes

Each line in the configuration file indicates one layer, and the network format for each layer to be defined in a separate file is "standard edges list", see Sec. A.5.3.

A.5.2 Non-edge-colored networks

If the multilayer is not edge-colored (i.e., inter-links are allowed), only one line is specified in the configuration file, with format:

`path_multilayer;path_to_layers_info;path_to_layers_layout`

where:

Variable	Type	Mandatory	Description
`path_multilayer`	String	Yes	Specify the path and the filename to the extended edges list to be used
`path_to_layers_info`	String	Yes	Specify the path and the filename to the file containing information about layers
`path_to_layers_layout`	String	Yes	Specify the path and the filename to the file containing information about nodes

In this case the network format required is "extended edges list", consisting of a single file. See Sec. A.5.4.

A.5.3 Standard edges list

A typical edges list is expected to be a file with at most three columns, giving the list of edges from a node (first column) to other nodes (second column), possibly weighted by an integer or floating number (third column). For instance:

```
1 2 0.5
1 3 1.4
...
18 124 0.1
```

is a typical weighted edges list.

IDs of nodes are expected to be sequential integers (starting from 0 or 1, up to the number of nodes in the network). Nevertheless, it is possible to import label-based edges list, where the IDs of nodes are labels (arbitrary integers or strings): in this case, one should check the appropriate box before importing the networks, to let **muxViz** know how to interpret the format. The edges list should follow the format:

```
alice bob 0.5
alice charlie 1.4
...
john david 0.1
```

In this specific case, it is mandatory to provide a layout file (see Sec. A.5.5) reporting each node label (field `nodeLabel`). This should look like:

```
nodeLabel
alice
bob
john
david
...
```

A.5.4 Extended edges list

An extended edges list is a new format that allows to specify all possible types of links, intra- and inter-layer. Each line specifies the source node (first column) and the source layer (second column), the destination node (third column) and the destination layer (fourth column), possibly weighted by an integer or floating number (fifth column). For instance:

```
1 1 2 1 0.5
1 1 3 1 1.4
...
18 2 124 2 0.1
```

is a typical weighted extended edges list. For label-based extended edges lists, the same rules of the standard edges lists apply.

A.5.5 Format of a layout file

The first line of the file must specify the name of the corresponding node attributes. Allowed attributes are:

The order of the columns should not be relevant. If `nodeLat` and `nodeLong` are specified, they will be automatically converted to Cartesian coordinates (through Mercator projection).

The properties of each node in the multilayer must be specified or default values will be used (i.e., automatic labeling and layouting). If the number of nodes in the network is different from the number of nodes provided in the layout file, it will be assumed that something is wrong with the layout file and default values will be used.

Variable	Type	Mandatory	Description
nodeID	Integer	Yes	ID to identify each node
nodeLabel	String	No	Label attribute
nodeX	Float	No	Cartesian coordinate x for the layout
nodeY	Float	No	Cartesian coordinate y for the layout
nodeLat	Float	No	Latitude for the geographic layout
nodeLong	Float	No	Longitude for the geographic layout

A.5.6 Format of a layer-info file

The first line of the file must specify the name of the corresponding layer attributes. Allowed attributes are:

Variable	Type	Mandatory	Description
layerID	Integer	Yes	ID to identify each layer
layerLabel	String	No	Label attribute

The order of the columns should not be relevant.

A.5.7 Format of a timeline file

This module allows to build nice animated visualizations corresponding to dynamical processes on the top of a multilayer network. For instance, one can visualize the movements of one (or more) individual(s) or random walker(s) in the network, the spreading of an epidemics or a meme in a social network, traffic and possible congestions in a transport or communication network, so forth so on.

The idea is to feed the module with a 'timeline' file where the change of the state of nodes and edges in the multilayer network are specified at each time step. The 'state' of an object can be altered by changing its color and/or its size. For instance, in the case of an epidemics spreading in a country, the size of each node (e.g., a meta-population describing a city) can be proportional to the population and the color can encode the amount of infected people. This description allows a wide variety of dynamics to be represented and visualized: for instance, setting the size of nodes and edges to zero when required, it is possible to visualize a time-varying multilayer network where nodes and edges appear or disappear over time.

The first line of the file must specify the name of the corresponding timeline attributes. Allowed attributes are:

The order of the columns is not relevant. If the network has L layers and you want to include the aggregate network in the visualization, then use $L+1$ in the layerID field for it.

Variable	Type	Mandatory	Description
`timeStep`	Integer	Yes	ID to identify time steps
`labelStep`	String	Yes	Snapshot label
`entity`	String	Yes	Object to modify, can be 'node' or 'edge'
`layerID`	Integer	Yes	ID to identify layers
`nodeID`	Integer or String	Yes	ID. Integer if entity is 'node' and String (e.g., '3-7', corresponding to the link from node 3 to node 7) if entity is 'edge'
`color`	String	Yes	Hex color (e.g. 11DADA) for entity
`sizeFactor`	Float	Yes	Relative size of the entity, scaling with respect to **muxViz** default size

References

[1] Artime, O. & al, e. *Multilayer Network Science. Theory and Applications: from Cells to Societies* (Cambridge University Press, 2021).

[2] Bianconi, G. *Multilayer networks: structure and function* (Oxford University Press, 2018).

[3] Zachary, W. W. An information flow model for conflict and fission in small groups. *Journal of Anthropological Research* **33**, 452–473 (1977).

[4] De Domenico, M. Multilayer Networks Illustrated (2020). Download from http://doi.org/10.17605/OSF.IO/GY53Khttp://doi.org/10.17605/OSF.IO/GY53K; Accessed: 2020-11-25.

[5] Ferrara, E., Varol, O., Davis, C., Menczer, F. & Flammini, A. The rise of social bots. *Communications of the ACM* **59**, 96–104 (2016).

[6] Stella, M., Ferrara, E. & De Domenico, M. Bots increase exposure to negative and inflammatory content in online social systems. *Proceedings of the National Academy of Sciences* **115**, 12435–12440 (2018).

[7] Kolda, T. G. & Bader, B. W. Tensor decompositions and applications. *SIAM Review* **51**, 455–500 (2009).

[8] Baggio, J. A. *et al.* Multiplex social ecological network analysis reveals how social changes affect community robustness more than resource depletion. *PNAS* **113**, 13708–13713 (2016).

[9] Timóteo, S., Correia, M., Rodríguez-Echeverría, S., Freitas, H. & Heleno, R. Multilayer networks reveal the spatial structure of seed-dispersal interactions across the great rift landscapes. *Nature Communications* **9**, 140 (2018).

[10] De Domenico, M., Solé-Ribalta, A., Omodei, E., Gómez, S. & Arenas, A. Ranking in interconnected multilayer networks reveals versatile nodes. *Nature Communications* **6**, 6868 (2015).

[11] Seidman, S. B. Network structure and minimum degree. *Social networks* **5**, 269–287 (1983).

[12] De Domenico, M., Lima, A., Mougel, P. & Musolesi, M. The anatomy of a scientific rumor. *Scientific reports* **3**, 2980 (2013).

[13] Rosvall, M. & Bergstrom, C. T. Maps of random walks on complex networks reveal community structure. *PNAS* **105**, 1118–1123 (2008).

[14] Edler, D., Bohlin, L. & Rosvall, M. Mapping higher-order network flows in memory and multilayer networks with infomap. *Algorithms* **10**, 112 (2017).

[15] De Domenico, M., Lancichinetti, A., Arenas, A. & Rosvall, M. Identifying modular flows on multilayer networks reveals highly overlapping organization in interconnected systems. *Physical Review X* **5**, 011027 (2015).

[16] Mangioni, G., Jurman, G. & DeDomenico, M. Multilayer flows in molecular networks identify biological modules in the human proteome. *IEEE Transactions on Network Science and Engineering* **7**, 411 (2018).

[17] Ghavasieh, A. & De Domenico, M. Enhancing transport properties in interconnected systems without altering their structure. *Physical Review Research* **2**, 013155 (2020).

M. De Domenico, *Multilayer Networks: Analysis and Visualization*, https://doi.org/10.1007/978-3-030-75718-2

[18] De Domenico, M., Solé-Ribalta, A., Gómez, S. & Arenas, A. Navigability of interconnected networks under random failures. *PNAS* **111**, 8351–8356 (2014).

[19] De Domenico, M., Porter, M. A. & Arenas, A. Muxviz: a tool for multilayer analysis and visualization of networks. *Journal of Complex Networks* **3**, 159–176 (2015).

[20] Watts, D. J. & Strogatz, S. H. Collective dynamics of small-world networks. *Nature* **393**, 440 (1998).

[21] Barabási, A.-L. & Albert, R. Emergence of scaling in random networks. *Science* **286**, 509–512 (1999).

[22] Barrat, A., Barthelemy, M. & Vespignani, A. *Dynamical processes on complex networks* (Cambridge University Press, 2008).

[23] Newman, M. *Networks: an introduction* (Oxford University Press, 2010).

[24] Barabási, A.-L. & Pósfai, M. *Network science* (Cambridge University Press, 2016).

[25] Caldarelli, G. *Scale-free networks: complex webs in nature and technology* (Oxford University Press, 2007).

[26] Dorogovtsev, S. N. *Lectures on complex networks*, vol. 24 (Oxford University Press, 2010).

[27] Estrada, E. *The structure of complex networks: theory and applications* (Oxford University Press, 2012).

[28] Latora, V., Nicosia, V. & Russo, G. *Complex networks: principles, methods and applications* (Cambridge University Press, 2017).

[29] Borgatti, S. P., Mehra, A., Brass, D. J. & Labianca, G. Network analysis in the social sciences. *science* **323**, 892–895 (2009).

[30] Jackson, M. O. Chapter 12 - an overview of social networks and economic applications*. vol. 1 of *Handbook of Social Economics*, 511 – 585 (North-Holland, 2011).

[31] Borgatti, S. P. & Halgin, D. S. On network theory. *Organization Science* **22**, 1168–1181 (2011).

[32] Sporns, O., Chialvo, D. R., Kaiser, M. & Hilgetag, C. C. Organization, development and function of complex brain networks. *Trends in Cognitive Sciences* **8**, 418–425 (2004).

[33] Bullmore, E. & Sporns, O. Complex brain networks: graph theoretical analysis of structural and functional systems. *Nature Reviews Neuroscience* **10**, 186 (2009).

[34] Meunier, D., Lambiotte, R. & Bullmore, E. T. Modular and hierarchically modular organization of brain networks. *Frontiers in Neuroscience* **4**, 200 (2010).

[35] Sporns, O. The human connectome: a complex network. *Annals of the New York Academy of Sciences* **1224**, 109–125 (2011).

[36] Bullmore, E. & Sporns, O. The economy of brain network organization. *Nature Reviews Neuroscience* **13**, 336 (2012).

[37] Sporns, O. Contributions and challenges for network models in cognitive neuroscience. *Nature Neuroscience* **17**, 652 (2014).

[38] Fallani, F. D. V., Richiardi, J., Chavez, M. & Achard, S. Graph analysis of functional brain networks: practical issues in translational neuroscience. *Phil. Trans. R. Soc. B* **369**, 20130521 (2014).

[39] Medaglia, J. D., Lynall, M.-E. & Bassett, D. S. Cognitive network neuroscience. *Journal of Cognitive Neuroscience* **27**, 1471–1491 (2015).

[40] Bassett, D. S. & Sporns, O. Network neuroscience. *Nature Neuroscience* **20**, 353 (2017).

[41] Yarden, Y. & Sliwkowski, M. X. Untangling the erbb signalling network. *Nature Reviews Molecular Cell biology* **2**, 127 (2001).

[42] Tyson, J. J., Chen, K. & Novak, B. Network dynamics and cell physiology. *Nature Reviews Molecular Cell Biology* **2**, 908 (2001).

[43] Kitano, H. Computational systems biology. *Nature* **420**, 206 (2002).

[44] Barabasi, A.-L. & Oltvai, Z. N. Network biology: understanding the cell's functional organization. *Nature Reviews Genetics* **5**, 101 (2004).
[45] Kitano, H. Biological robustness. *Nature Reviews Genetics* **5**, 826 (2004).
[46] Thiery, J. P. & Sleeman, J. P. Complex networks orchestrate epithelial–mesenchymal transitions. *Nature Reviews Molecular Cell biology* **7**, 131 (2006).
[47] Sharan, R. & Ideker, T. Modeling cellular machinery through biological network comparison. *Nature Biotechnology* **24**, 427 (2006).
[48] Barabási, A.-L., Gulbahce, N. & Loscalzo, J. Network medicine: a network-based approach to human disease. *Nature Reviews Genetics* **12**, 56 (2011).
[49] Strogatz, S. H. Exploring complex networks. *nature* **410**, 268 (2001).
[50] Albert, R. & Barabási, A.-L. Statistical mechanics of complex networks. *Reviews of Modern Physics* **74**, 47 (2002).
[51] Newman, M. E. The structure and function of complex networks. *SIAM review* **45**, 167–256 (2003).
[52] Boccaletti, S., Latora, V., Moreno, Y., Chavez, M. & Hwang, D.-U. Complex networks: Structure and dynamics. *Physics reports* **424**, 175–308 (2006).
[53] Arenas, A., Díaz-Guilera, A., Kurths, J., Moreno, Y. & Zhou, C. Synchronization in complex networks. *Physics Reports* **469**, 93–153 (2008).
[54] Dorogovtsev, S. N., Goltsev, A. V. & Mendes, J. F. Critical phenomena in complex networks. *Reviews of Modern Physics* **80**, 1275 (2008).
[55] Newman, M. E. Communities, modules and large-scale structure in networks. *Nature Physics* **8**, 25 (2012).
[56] Holme, P. & Saramäki, J. Temporal networks. *Physics reports* **519**, 97–125 (2012).
[57] Holme, P. Modern temporal network theory: a colloquium. *The European Physical Journal B* **88**, 234 (2015).
[58] D'Souza, R. M. & Nagler, J. Anomalous critical and supercritical phenomena in explosive percolation. *Nature Physics* **11**, 531 (2015).
[59] Pastor-Satorras, R., Castellano, C., Van Mieghem, P. & Vespignani, A. Epidemic processes in complex networks. *Reviews of Modern Physics* **87**, 925 (2015).
[60] Fortunato, S. & Hric, D. Community detection in networks: A user guide. *Physics Reports* **659**, 1–44 (2016).
[61] Wang, Z. *et al.* Statistical physics of vaccination. *Physics Reports* **664**, 1–113 (2016).
[62] Montoya, J. M., Pimm, S. L. & Solé, R. V. Ecological networks and their fragility. *Nature* **442**, 259 (2006).
[63] Ings, T. C. *et al.* Ecological networks–beyond food webs. *Journal of Animal Ecology* **78**, 253–269 (2009).
[64] Erdös, P. & Rényi, A. On random graphs, i. *Publicationes Mathematicae (Debrecen)* **6**, 290–297 (1959).
[65] Newman, M. E., Watts, D. J. & Strogatz, S. H. Random graph models of social networks. *PNAS* **99**, 2566–2572 (2002).
[66] Ravasz, E. & Barabási, A.-L. Hierarchical organization in complex networks. *Physical Review E* **67**, 026112 (2003).
[67] Sales-Pardo, M., Guimera, R., Moreira, A. A. & Amaral, L. A. N. Extracting the hierarchical organization of complex systems. *PNAS* **104**, 15224–15229 (2007).
[68] Clauset, A., Moore, C. & Newman, M. E. Hierarchical structure and the prediction of missing links in networks. *Nature* **453**, 98 (2008).
[69] Peixoto, T. P. Hierarchical block structures and high-resolution model selection in large networks. *Physical Review X* **4**, 011047 (2014).

[70] Corominas-Murtra, B., Goñi, J., Solé, R. V. & Rodríguez-Caso, C. On the origins of hierarchy in complex networks. *PNAS* **110**, 13316–13321 (2013).
[71] Girvan, M. & Newman, M. E. Community structure in social and biological networks. *PNAS* **99**, 7821–7826 (2002).
[72] Donetti, L. & Munoz, M. A. Detecting network communities: a new systematic and efficient algorithm. *Journal of Statistical Mechanics* **2004**, P10012 (2004).
[73] Duch, J. & Arenas, A. Community detection in complex networks using extremal optimization. *Physical review E* **72**, 027104 (2005).
[74] Newman, M. E. Modularity and community structure in networks. *PNAS* **103**, 8577–8582 (2006).
[75] Reichardt, J. & Bornholdt, S. Statistical mechanics of community detection. *Physical Review E* **74**, 016110 (2006).
[76] Fortunato, S. & Barthelemy, M. Resolution limit in community detection. *PNAS* **104**, 36–41 (2007).
[77] Arenas, A., Fernandez, A. & Gomez, S. Analysis of the structure of complex networks at different resolution levels. *New Journal of Physics* **10**, 053039 (2008).
[78] Blondel, V. D., Guillaume, J.-L., Lambiotte, R. & Lefebvre, E. Fast unfolding of communities in large networks. *Journal of Statistical Mechanics* **2008**, P10008 (2008).
[79] Lancichinetti, A. & Fortunato, S. Community detection algorithms: a comparative analysis. *Physical Review E* **80**, 056117 (2009).
[80] Lancichinetti, A., Fortunato, S. & Kertész, J. Detecting the overlapping and hierarchical community structure in complex networks. *New Journal of Physics* **11**, 033015 (2009).
[81] Peixoto, T. P. Parsimonious module inference in large networks. *Physical Review Letters* **110**, 148701 (2013).
[82] Newman, M. E. & Peixoto, T. P. Generalized communities in networks. *Physical Review Letters* **115**, 088701 (2015).
[83] Peixoto, T. P. Model selection and hypothesis testing for large-scale network models with overlapping groups. *Physical Review X* **5**, 011033 (2015).
[84] Guimera, R. & Amaral, L. A. N. Functional cartography of complex metabolic networks. *Nature* **433**, 895 (2005).
[85] Pons, P. & Latapy, M. Computing communities in large networks using random walks. *J. Graph Algorithms Appl.* **10**, 191–218 (2006).
[86] Rosvall, M. & Bergstrom, C. T. An information-theoretic framework for resolving community structure in complex networks. *PNAS* **104**, 7327–7331 (2007).
[87] Mucha, P. J., Richardson, T., Macon, K., Porter, M. A. & Onnela, J.-P. Community structure in time-dependent, multiscale, and multiplex networks. *Science* **328**, 876–878 (2010).
[88] Lambiotte, R., Delvenne, J.-C. & Barahona, M. Random walks, markov processes and the multiscale modular organization of complex networks. *IEEE Transactions on Network Science and Engineering* **1**, 76–90 (2014).
[89] De Domenico, M. Diffusion geometry unravels the emergence of functional clusters in collective phenomena. *Physical Review Letters* **118**, 168301 (2017).
[90] Peixoto, T. P. & Rosvall, M. Modelling sequences and temporal networks with dynamic community structures. *Nature Communications* **8**, 582 (2017).
[91] Dorogovtsev, S. N., Mendes, J. F. F. & Samukhin, A. N. Structure of growing networks with preferential linking. *Physical Review Letters* **85**, 4633 (2000).

[92] Bianconi, G. & Barabási, A.-L. Bose-einstein condensation in complex networks. *Physical Review Letters* **86**, 5632 (2001).
[93] Krapivsky, P. L. & Redner, S. Organization of growing random networks. *Physical Review E* **63**, 066123 (2001).
[94] Bianconi, G. & Barabási, A.-L. Competition and multiscaling in evolving networks. *EuroPhysics Letters* **54**, 436 (2001).
[95] Callaway, D. S., Hopcroft, J. E., Kleinberg, J. M., Newman, M. E. & Strogatz, S. H. Are randomly grown graphs really random? *Physical Review E* **64**, 041902 (2001).
[96] Caldarelli, G., Capocci, A., De Los Rios, P. & Munoz, M. A. Scale-free networks from varying vertex intrinsic fitness. *Physical Review Letters* **89**, 258702 (2002).
[97] De Domenico, M. & Arenas, A. Modeling structure and resilience of the dark network. *Physical Review E* **95**, 022313 (2017).
[98] Molloy, M. & Reed, B. A critical point for random graphs with a given degree sequence. *Random structures & algorithms* **6**, 161–180 (1995).
[99] Park, J. & Newman, M. E. Statistical mechanics of networks. *Physical Review E* **70**, 066117 (2004).
[100] Robins, G., Pattison, P., Kalish, Y. & Lusher, D. An introduction to exponential random graph (p*) models for social networks. *Social Networks* **29**, 173–191 (2007).
[101] Robins, G., Snijders, T., Wang, P., Handcock, M. & Pattison, P. Recent developments in exponential random graph (p*) models for social networks. *Social Networks* **29**, 192–215 (2007).
[102] Wang, P., Robins, G., Pattison, P. & Lazega, E. Exponential random graph models for multilevel networks. *Social Networks* **35**, 96–115 (2013).
[103] Albert, R., Jeong, H. & Barabási, A.-L. Error and attack tolerance of complex networks. *Nature* **406**, 378 (2000).
[104] Callaway, D. S., Newman, M. E., Strogatz, S. H. & Watts, D. J. Network robustness and fragility: Percolation on random graphs. *Physical Review Letters* **85**, 5468 (2000).
[105] Cohen, R., Erez, K., Ben-Avraham, D. & Havlin, S. Resilience of the internet to random breakdowns. *Physical Review Letters* **85**, 4626 (2000).
[106] Cohen, R., Erez, K., Ben-Avraham, D. & Havlin, S. Breakdown of the internet under intentional attack. *Physical Review Letters* **86**, 3682 (2001).
[107] Carreras, B. A., Lynch, V. E., Dobson, I. & Newman, D. E. Critical points and transitions in an electric power transmission model for cascading failure blackouts. *Chaos: An interdisciplinary journal of nonlinear science* **12**, 985–994 (2002).
[108] Watts, D. J. A simple model of global cascades on random networks. *PNAS* **99**, 5766–5771 (2002).
[109] Holme, P., Kim, B. J., Yoon, C. N. & Han, S. K. Attack vulnerability of complex networks. *Physical review E* **65**, 056109 (2002).
[110] Motter, A. E. & Lai, Y.-C. Cascade-based attacks on complex networks. *Physical Review E* **66**, 065102 (2002).
[111] Crucitti, P., Latora, V. & Marchiori, M. A topological analysis of the italian electric power grid. *Physica A: Statistical mechanics and its applications* **338**, 92–97 (2004).
[112] Crucitti, P., Latora, V., Marchiori, M. & Rapisarda, A. Error and attack tolerance of complex networks. *Physica A: Statistical mechanics and its applications* **340**, 388–394 (2004).
[113] Zhao, L., Park, K. & Lai, Y.-C. Attack vulnerability of scale-free networks due to cascading breakdown. *Physical Review E* **70**, 035101 (2004).
[114] Crucitti, P., Latora, V. & Marchiori, M. Model for cascading failures in complex networks. *Physical Review E* **69**, 045104 (2004).
[115] Motter, A. E. Cascade control and defense in complex networks. *Physical Review Letters* **93**, 098701 (2004).

[116] Kinney, R., Crucitti, P., Albert, R. & Latora, V. Modeling cascading failures in the north american power grid. *The European Physical Journal B* **46**, 101–107 (2005).
[117] Smart, A. G., Amaral, L. A. & Ottino, J. M. Cascading failure and robustness in metabolic networks. *PNAS* **105**, 13223–13228 (2008).
[118] Dueñas-Osorio, L. & Vemuru, S. M. Cascading failures in complex infrastructure systems. *Structural Safety* **31**, 157–167 (2009).
[119] Wang, J.-W. & Rong, L.-L. Cascade-based attack vulnerability on the us power grid. *Safety Science* **47**, 1332–1336 (2009).
[120] Pahwa, S., Scoglio, C. & Scala, A. Abruptness of cascade failures in power grids. *Scientific Reports* **4**, 3694 (2014).
[121] Braunstein, A., Dall?Asta, L., Semerjian, G. & Zdeborová, L. Network dismantling. *PNAS* **113**, 12368–12373 (2016).
[122] Kitsak, M. *et al.* Identification of influential spreaders in complex networks. *Nature Physics* **6**, 888 (2010).
[123] Morone, F. & Makse, H. A. Influence maximization in complex networks through optimal percolation. *Nature* **524**, 65 (2015).
[124] Guimerà, R. & Sales-Pardo, M. Missing and spurious interactions and the reconstruction of complex networks. *PNAS* **106**, 22073–22078 (2009).
[125] Newman, M. E. & Clauset, A. Structure and inference in annotated networks. *Nature Communications* **7**, 11863 (2016).
[126] Hric, D., Peixoto, T. P. & Fortunato, S. Network structure, metadata, and the prediction of missing nodes and annotations. *Physical Review X* **6**, 031038 (2016).
[127] Peel, L., Larremore, D. B. & Clauset, A. The ground truth about metadata and community detection in networks. *Science Advances* **3**, e1602548 (2017).
[128] Ford, L. R. & Fulkerson, D. R. Maximal flow through a network. *Canadian Journal of Mathematics* **8**, 399–404 (1956).
[129] De Domenico, M. *et al.* Mathematical formulation of multilayer networks. *Physical Review X* **3**, 041022 (2013).
[130] Capra, F. & Luisi, P. L. *The systems view of life: A unifying vision* (Cambridge University Press, 2014).
[131] Kapferer, B. Norms and the manipulation of relationships in a work context. In Michell, J. (ed.) *Social Networks in Urban Situations* (Manchester University Press, Manchester, 1969).
[132] Granovetter, M. S. The strength of weak ties. *American Journal of Sociology* **78**, 1360–1380 (1973).
[133] Gomez, S. *et al.* Diffusion dynamics on multiplex networks. *Physical Review Letters* **110**, 028701 (2013).
[134] Verbrugge, L. M. Multiplexity in adult friendships. *Social Forces* **57**, 1286–1309 (1979).
[135] Padgett, J. F. & Ansell, C. K. Robust action and the rise of the medici, 1400-1434. *American journal of sociology* **98**, 1259–1319 (1993).
[136] Padgett, J. F. Marriage and elite structure in renaissance florence, 1282-1500. In *Proceedings of Social Science History Association Annual Meeting*, 1 (Social Science History Association, 1994).
[137] Jacob, F. The logic of living systems: a history of heredity (Lane, 1974).
[138] Willner, S. N., Otto, C. & Levermann, A. Global economic response to river floods. *Nature Climate Change* 1 (2018).
[139] Buldyrev, S. V., Parshani, R., Paul, G., Stanley, H. E. & Havlin, S. Catastrophic cascade of failures in interdependent networks. *Nature* **464**, 1025 (2010).
[140] Vespignani, A. Complex networks: The fragility of interdependency. *Nature* **464**, 984 (2010).
[141] Gao, J., Buldyrev, S. V., Stanley, H. E. & Havlin, S. Networks formed from interdependent networks. *Nature Physics* **8**, 40 (2012).

[142] Boccaletti, S. *et al.* The structure and dynamics of multilayer networks. *Physics Reports* **544**, 1–122 (2014).

[143] Kivelä, M. *et al.* Multilayer networks. *Journal of Complex Networks* **2**, 203–271 (2014).

[144] Lee, K.-M., Min, B. & Goh, K.-I. Towards real-world complexity: an introduction to multiplex networks. *The European Physical Journal B* **88**, 48 (2015).

[145] De Domenico, M., Granell, C., Porter, M. A. & Arenas, A. The physics of spreading processes in multilayer networks. *Nature Physics* **12**, 901 (2016).

[146] De Domenico, M. Multilayer modeling and analysis of human brain networks. *GigaScience* **6**, 1–8 (2017).

[147] De Domenico, M. Multilayer network modeling of integrated biological systems. Comment on "Network science of biological systems at different scales: A review" by Gosak et al. *Physics of Life Reviews* (2018).

[148] Cardillo, A. *et al.* Emergence of network features from multiplexity. *Scientific Reports* **3** (2013).

[149] Nicosia, V., Bianconi, G., Latora, V. & Barthelemy, M. Growing multiplex networks. *Physical Review Letters* **111**, 058701 (2013).

[150] Bianconi, G. Statistical mechanics of multiplex networks: Entropy and overlap. *Physical Review E* **87**, 062806 (2013).

[151] Battiston, F., Nicosia, V. & Latora, V. Structural measures for multiplex networks. *Physical Review E* **89**, 032804 (2014).

[152] Nicosia, V. & Latora, V. Measuring and modeling correlations in multiplex networks. *Physical Review E* **92**, 032805 (2015).

[153] Radicchi, F. & Bianconi, G. Redundant interdependencies boost the robustness of multiplex networks. *Physical Review X* **7**, 011013 (2017).

[154] Pilosof, S., Porter, M. A., Pascual, M. & Kéfi, S. The multilayer nature of ecological networks. *Nature Ecology & Evolution* **1**, 0101 (2017).

[155] Silk, M. J., Finn, K. R., Porter, M. A. & Pinter-Wollman, N. Can multilayer networks advance animal behavior research? *Trends in ecology & evolution* **33**, 376–378 (2018).

[156] Stella, M., Selakovic, S., Antonioni, A. & Andreazzi, C. Ecological multiplex interactions determine the role of species for parasite spread amplification. *eLife* **7** (2018).

[157] Arenas, A. & De Domenico, M. Nonlinear dynamics on interconnected networks. *Physica D: Nonlinear Phenomena* **323**, 1–4 (2016).

[158] Sole-Ribalta, A. *et al.* Spectral properties of the laplacian of multiplex networks. *Physical Review E* **88**, 032807 (2013).

[159] Requejo, R. J. & Díaz-Guilera, A. Replicator dynamics with diffusion on multiplex networks. *Physical Review E* **94**, 022301 (2016).

[160] Brechtel, A., Gramlich, P., Ritterskamp, D., Drossel, B. & Gross, T. Master stability functions reveal diffusion-driven pattern formation in networks. *Physical Review E* **97** (2018).

[161] Tejedor, A., Longjas, A., Foufoula-Georgiou, E., Georgiou, T. T. & Moreno, Y. Diffusion dynamics and optimal coupling in multiplex networks with directed layers. *Physical Review X* **8**, 031071 (2018).

[162] Solé-Ribalta, A., De Domenico, M., Gómez, S. & Arenas, A. Random walk centrality in interconnected multilayer networks. *Physica D: Nonlinear Phenomena* **323**, 73–79 (2016).

[163] Lacasa, L. *et al.* Multiplex Decomposition of Non-Markovian Dynamics and the Hidden Layer Reconstruction Problem. *Physical Review X* **8**, 031038 (2018).

[164] Valdeolivas, A. *et al.* Random walk with restart on multiplex and heterogeneous biological networks. *Bioinformatics* (2018).

[165] Singh, A., Ghosh, S., Jalan, S. & Kurths, J. Synchronization in delayed multiplex networks. *EPL (Europhysics Letters)* **111**, 30010 (2015).

[166] Skardal, P. S. & Arenas, A. Control of coupled oscillator networks with application to microgrid technologies. *Science Advances* **1**, e1500339 (2015).
[167] Zhang, X., Boccaletti, S., Guan, S. & Liu, Z. Explosive synchronization in adaptive and multilayer networks. *Physical Review Letters* **114**, 038701 (2015).
[168] Saa, A. Symmetries and synchronization in multilayer random networks. *Physical Review E* **97**, 042304 (2018).
[169] Leyva, I. *et al.* Relay synchronization in multiplex networks. *Scientific reports* **8** (2018).
[170] Yuan, Z., Zhao, C., Wang, W.-X., Di, Z. & Lai, Y.-C. Exact controllability of multiplex networks. *New Journal of Physics* **16**, 103036 (2014).
[171] Pósfai, M., Gao, J., Cornelius, S. P., Barabási, A.-L. & D'Souza, R. M. Controllability of multiplex, multi-time-scale networks. *Physical Review E* **94**, 032316 (2016).
[172] Gómez-Gardenes, J., Reinares, I., Arenas, A. & Floría, L. M. Evolution of cooperation in multiplex networks. *Scientific Reports* **2**, 620 (2012).
[173] Matamalas, J. T., Poncela-Casasnovas, J., Gómez, S. & Arenas, A. Strategical incoherence regulates cooperation in social dilemmas on multiplex networks. *Scientific Reports* **5**, 9519 (2015).
[174] Battiston, F., Perc, M. & Latora, V. Determinants of public cooperation in multiplex networks. *New Journal of Physics* **19**, 073017 (2017).
[175] Estrada, E. & Gómez-Gardeñes, J. Communicability reveals a transition to coordinated behavior in multiplex networks. *Physical Review E* **89**, 042819 (2014).
[176] Wang, H. *et al.* Effect of the interconnected network structure on the epidemic threshold. *Physical Review E* **88**, 022801 (2013).
[177] Sahneh, F. D., Scoglio, C. & Van Mieghem, P. Generalized epidemic mean-field model for spreading processes over multilayer complex networks. *IEEE/ACM Transactions on Networking (TON)* **21**, 1609–1620 (2013).
[178] Buono, C., Alvarez-Zuzek, L. G., Macri, P. A. & Braunstein, L. A. Epidemics in partially overlapped multiplex networks. *PloS one* **9**, e92200 (2014).
[179] Valdano, E., Ferreri, L., Poletto, C. & Colizza, V. Analytical computation of the epidemic threshold on temporal networks. *Physical Review X* **5**, 021005 (2015).
[180] Bianconi, G. Epidemic spreading and bond percolation on multilayer networks. *Journal of Statistical Mechanics: Theory and Experiment* **2017**, 034001 (2017).
[181] Tan, F., Wu, J., Xia, Y. & Chi, K. T. Traffic congestion in interconnected complex networks. *Physical Review E* **89**, 062813 (2014).
[182] Solé-Ribalta, A., Gómez, S. & Arenas, A. Congestion Induced by the Structure of Multiplex Networks. *Physical Review Letters* **116**, 108701 (2016).
[183] Chodrow, P. S., Al-Awwad, Z., Jiang, S. & González, M. C. Demand and Congestion in Multiplex Transportation Networks. *PloS one* **11**, e0161738 (2016).
[184] Jang, S., Lee, J. S., Hwang, S. & Kahng, B. Ashkin-Teller model and diverse opinion phase transitions on multiplex networks. *Physical Review E* **92**, 022110 (2015).
[185] Diakonova, M., Nicosia, V., Latora, V. & Miguel, M. S. Irreducibility of multilayer network dynamics: the case of the voter model. *New Journal of Physics* **18**, 023010 (2016).
[186] Artime, O., Fernández-Gracia, J., Ramasco, J. J. & San Miguel, M. Joint effect of ageing and multilayer structure prevents ordering in the voter model. *Scientific Reports* **7**, 7166 (2017).

[187] Antonopoulos, C. G. & Shang, Y. Opinion formation in multiplex networks with general initial distributions. *Scientific Reports* **8**, 2852 (2018).
[188] Yagan, O. & Gligor, V. Analysis of complex contagions in random multiplex networks. *Physical Review E* **86**, 036103 (2012).
[189] Hu, Y., Havlin, S. & Makse, H. a. Conditions for Viral Influence Spreading through Multiplex Correlated Social Networks. *Physical Review X* **4**, 021031 (2014).
[190] Ramezanian, R., Magnani, M., Salehi, M. & Montesi, D. Diffusion of innovations over multiplex social networks. In *Artificial Intelligence and Signal Processing (AISP), 2015 International Symposium on*, 300–304 (IEEE, 2015).
[191] Traag, V. A. Complex contagion of campaign donations. *PloS one* **11**, e0153539 (2016).
[192] Wang, X. *et al.* Promoting information diffusion through interlayer recovery processes in multiplex networks. *Physical Review E* **96**, 032304 (2017).
[193] Radicchi, F. & Arenas, A. Abrupt transition in the structural formation of interconnected networks. *Nature Physics* **9**, 717 (2013).
[194] Dickison, M., Havlin, S. & Stanley, H. E. Epidemics on interconnected networks. *Physical Review E* **85**, 066109 (2012).
[195] Cozzo, E., Baños, R. A., Meloni, S. & Moreno, Y. Contact-based social contagion in multiplex networks. *Physical Review E* **88**, 050801 (2013).
[196] Sanz, J., Xia, C.-Y., Meloni, S. & Moreno, Y. Dynamics of Interacting Diseases. *Physical Review X* **4**, 041005 (2014).
[197] de Arruda, G. F., Cozzo, E., Peixoto, T. P., Rodrigues, F. A. & Moreno, Y. Disease Localization in Multilayer Networks. *Physical Review X* **7**, 011014 (2017).
[198] Danziger, M. M., Bonamassa, I., Boccaletti, S. & Havlin, S. Dynamic interdependence and competition in multilayer networks. *Nature Physics* **In press** (2018).
[199] Funk, S., Gilad, E., Watkins, C. & Jansen, V. A. A. The spread of awareness and its impact on epidemic outbreaks. *PNAS* **106**, 6872–6877 (2009).
[200] Wu, Q., Fu, X., Small, M. & Xu, X.-J. The impact of awareness on epidemic spreading in networks. *Chaos* **22**, 013101 (2012).
[201] Lima, A., De Domenico, M., Pejovic, V. & Musolesi, M. Exploiting cellular data for disease containment and information campaigns strategies in country-wide epidemics. In *Proceedings of Third International Conference on the Analysis of Mobile Phone Datasets, Boston, USA*, 1 (NETMOB, 2013).
[202] Granell, C., Gómez, S. & Arenas, A. Dynamical interplay between awareness and epidemic spreading in multiplex networks. *Physical Review Letters* **111** (2013).
[203] Massaro, E. & Bagnoli, F. Epidemic spreading and risk perception in multiplex networks: a self-organized percolation method. *Physical Review E* **90**, 052817 (2014).
[204] Lima, A., De Domenico, M., Pejovic, V. & Musolesi, M. Disease containment strategies based on mobility and information dissemination. *Scientific Reports* **5**, 10650 (2015).
[205] Funk, S. *et al.* Nine challenges in incorporating the dynamics of behaviour in infectious diseases models. *Epidemics* **10**, 21–25 (2015).
[206] Wang, Z., M. A. Andrews, Z.-X. W., Wang, L. & Bauch, C. T. Coupled disease–behavior dynamics on complex networks: A review. *Physics of Life Reviews* **15**, 1–29 (2015).
[207] Azimi-Tafreshi, N. Cooperative epidemics on multiplex networks. *Physical Review E* **93**, 042303 (2016).

[208] Velásquez-Rojas, F. Interacting opinion and disease dynamics in multiplex networks: Discontinuous phase transition and nonmonotonic consensus times. *Physical Review E* **95**, 052315 (2017).
[209] Czaplicka, A., Toral, R. & Miguel, M. S. Competition of simple and complex adoption on interdependent networks. *Physical Review E* **94** (2016).
[210] Amato, R., Díaz-Guilera, A. & Kleineberg, K.-K. Interplay between social influence and competitive strategical games in multiplex networks. *Scientific Reports* **7**, 7087 (2017).
[211] Soriano-Paños, D., Lotero, L., Arenas, A. & Gómez-Gardeñes, J. Spreading Processes in Multiplex Metapopulations Containing Different Mobility Networks. *Physical Review X* **8**, 031039 (2018).
[212] Nicosia, V., Skardal, P. S., Arenas, A. & Latora, V. Collective Phenomena Emerging from the Interactions between Dynamical Processes in Multiplex Networks. *Physical Review Letters* **118**, 138302 (2017).
[213] Gomez-Gardenes, J., de Domenico, M., Gutierrez, G., Arenas, A. & Gomez, S. Layer-layer competition in multiplex complex networks. *Phil. Trans. R. Soc. A* **373**, 20150117– (2015).
[214] Kim, J. Y. & Goh, K.-I. I. Coevolution and correlated multiplexity in multiplex networks. *Physical Review Letters* **111**, 058702 (2013).
[215] Nicosia, V., Bianconi, G., Latora, V. & Barthelemy, M. Nonlinear growth and condensation in multiplex networks. *Physical Review E* **90**, 042807 (2014).
[216] Santoro, A., Latora, V., Nicosia, G. & Nicosia, V. Pareto optimality in multilayer network growth. *Physical Review Letters* **121**, 128302 (2018).
[217] Rosato, V. *et al.* Modelling interdependent infrastructures using interacting dynamical models. *International Journal of Critical Infrastructures* **4**, 63–79 (2008).
[218] Parshani, R., Buldyrev, S. V. & Havlin, S. Interdependent Networks: Reducing the Coupling Strength Leads to a Change from a First to Second Order Percolation Transition. *Physical Review Letters* **105**, 048701 (2010).
[219] Gao, J., Buldyrev, S. V., Havlin, S. & Stanley, H. E. Robustness of a network of networks. *Physical Review Letters* **107** (2011).
[220] Valdez, L. D., Macri, P. A., Stanley, H. E. & Braunstein, L. A. Triple point in correlated interdependent networks. *Physical Review E* **88**, 050803 (2013).
[221] Radicchi, F. Driving interconnected networks to supercriticality. *Physical Review X* **4**, 021014 (2014).
[222] Liu, X., Stanley, H. E. & Gao, J. Breakdown of interdependent directed networks. *PNAS* **113**, 1138–1143 (2016).
[223] Yuan, X., Hu, Y., Stanley, H. E. & Havlin, S. Eradicating catastrophic collapse in interdependent networks via reinforced nodes. *PNAS* **114**, 3311–3315 (2017).
[224] Zhang, Y., Arenas, A. & Yağan, O. Cascading failures in interdependent systems under a flow redistribution model. *Physical Review E* **97**, 022307 (2018).
[225] Brummitt, C. D., D?Souza, R. M. & Leicht, E. A. Suppressing cascades of load in interdependent networks. *PNAS* **109**, E680–E689 (2012).
[226] Baxter, G. J., Dorogovtsev, S. N., Goltsev, A. V. & Mendes, J. F. F. Avalanche Collapse of Interdependent Networks. *Physical Review Letters* **109**, 248701 (2012).
[227] Brummitt, C. D., Lee, K.-M. & Goh, K.-I. Multiplexity-facilitated cascades in networks. *Physical Review E* **85**, 045102 (2012).
[228] Bianconi, G., Dorogovtsev, S. N. & Mendes, J. F. Mutually connected component of networks of networks with replica nodes. *Physical Review E* **91**, 012804 (2015).

[229] Bianconi, G. & Dorogovtsev, S. N. Multiple percolation transitions in a configuration model of a network of networks. *Physical Review E* **89**, 062814 (2014).

[230] Radicchi, F. Percolation in real interdependent networks. *Nature Physics* **11**, 597 (2015).

[231] Hackett, A., Cellai, D., Gómez, S., Arenas, A. & Gleeson, J. P. Bond percolation on multiplex networks. *Physical Review X* **6**, 021002 (2016).

[232] Min, B., Yi, S. D., Lee, K.-M. & Goh, K.-I. Network robustness of multiplex networks with interlayer degree correlations. *Physical Review E* **89**, 042811 (2014).

[233] Cellai, D., López, E., Zhou, J., Gleeson, J. P. & Bianconi, G. Percolation in multiplex networks with overlap. *Physical Review E* **88**, 052811 (2013).

[234] Osat, S., Faqeeh, A. & Radicchi, F. Optimal percolation on multiplex networks. *Nature Communications* **8**, 1540 (2017).

[235] Baxter, G. J., Timár, G. & Mendes, J. F. F. Targeted damage to interdependent networks. *Phys. Rev. E* **98**, 032307 (2018).

[236] Majdandzic, A. *et al.* Multiple tipping points and optimal repairing in interacting networks. *Nature Communications* **7**, 10850 (2016).

[237] Singh, R. K. & Sinha, S. Optimal interdependence enhances the dynamical robustness of complex systems. *Physical Review E* **96** (2017).

[238] Peixoto, T. P. Inferring the mesoscale structure of layered, edge-valued, and time-varying networks. *Physical Review E* **92**, 042807 (2015).

[239] Bazzi, M., Jeub, L. G., Arenas, A., Howison, S. D. & Porter, M. A. A framework for the construction of generative models for mesoscale structure in multilayer networks. *Physical Review Research* **2**, 023100 (2020).

[240] Chung, F. R. & Graham, F. C. *Spectral graph theory.* 92 (American Mathematical Soc., 1997).

[241] Noh, J. D. & Rieger, H. Random walks on complex networks. *Physical Review Letters* **92**, 118701 (2004).

[242] Masuda, N., Porter, M. A. & Lambiotte, R. Random walks and diffusion on networks. *Physics reports* (2017).

[243] Wang, Z., Wang, L., Szolnoki, A. & Perc, M. Evolutionary games on multilayer networks: a colloquium. *The European physical journal B* **88**, 124 (2015).

[244] Golubitsky, M., Stewart, I. & Török, A. Patterns of synchrony in coupled cell networks with multiple arrows. *SIAM Journal on Applied Dynamical Systems* **4**, 78–100 (2005).

[245] Min, B., Do Yi, S., Lee, K.-M. & Goh, K.-I. Network robustness of multiplex networks with interlayer degree correlations. *Physical Review E* **89**, 042811 (2014).

[246] Reis, S. D. *et al.* Avoiding catastrophic failure in correlated networks of networks. *Nature Physics* **10**, 762 (2014).

[247] Gemmetto, V. & Garlaschelli, D. Multiplexity versus correlation: the role of local constraints in real multiplexes. *Scientific reports* **5**, 9120 (2015).

[248] Cozzo, E. *et al.* Structure of triadic relations in multiplex networks. *New Journal of Physics* **17**, 073029 (2015).

[249] Solá, L. *et al.* Eigenvector centrality of nodes in multiplex networks. *Chaos* **23**, 033131 (2013).

[250] Halu, A., Mondragón, R. J., Panzarasa, P. & Bianconi, G. Multiplex pagerank. *PloS one* **8**, e78293 (2013).

[251] Iacovacci, J., Rahmede, C., Arenas, A. & Bianconi, G. Functional multiplex pagerank. *EPL (Europhysics Letters)* **116**, 28004 (2016).

[252] Rahmede, C., Iacovacci, J., Arenas, A. & Bianconi, G. Centralities of nodes and influences of layers in large multiplex networks. *Journal of Complex Networks* **6**, 733–752 (2017).

[253] Taylor, D., Myers, S. A., Clauset, A., Porter, M. A. & Mucha, P. J. Eigenvector-based centrality measures for temporal networks. *Multiscale Modeling & Simulation* **15**, 537–574 (2017).

[254] Solé-Ribalta, A., De Domenico, M., Gómez, S. & Arenas, A. Centrality rankings in multiplex networks. In *Proceedings of the 2014 ACM conference on Web science*, 149–155 (ACM, 2014).

[255] Brin, S. & Page, L. The anatomy of a large-scale hypertextual web search engine. *Computer networks and ISDN systems* **30**, 107–117 (1998).

[256] Carmi, S., Havlin, S., Kirkpatrick, S., Shavitt, Y. & Shir, E. A model of internet topology using k-shell decomposition. *PNAS* **104**, 11150–11154 (2007).

[257] Azimi-Tafreshi, N., Gómez-Gardenes, J. & Dorogovtsev, S. k- core percolation on multiplex networks. *Physical Review E* **90**, 032816 (2014).

[258] Freeman, L. C. Centrality in social networks conceptual clarification. *Social networks* **1**, 215–239 (1978).

[259] Opsahl, T., Agneessens, F. & Skvoretz, J. Node centrality in weighted networks: Generalizing degree and shortest paths. *Social networks* **32**, 245–251 (2010).

[260] Kalir, S. *et al.* Ordering genes in a flagella pathway by analysis of expression kinetics from living bacteria. *Science* **292**, 2080–2083 (2001).

[261] Milo, R. *et al.* Network motifs: simple building blocks of complex networks. *Science* **298**, 824–827 (2002).

[262] Yeger-Lotem, E. *et al.* Network motifs in integrated cellular networks of transcription–regulation and protein–protein interaction. *PNAS* **101**, 5934–5939 (2004).

[263] Milo, R. *et al.* Superfamilies of evolved and designed networks. *Science* **303**, 1538–1542 (2004).

[264] Wernicke, S. & Rasche, F. Fanmod: a tool for fast network motif detection. *Bioinformatics* **22**, 1152–1153 (2006).

[265] Kossinets, G. & Watts, D. J. Empirical analysis of an evolving social network. *science* **311**, 88–90 (2006).

[266] Grassberger, P. Percolation transitions in the survival of interdependent agents on multiplex networks, catastrophic cascades, and solid-on-solid surface growth. *Physical Review E* **91**, 062806 (2015).

[267] Stella, M., Beckage, N. M., Brede, M. & De Domenico, M. Multiplex model of mental lexicon reveals explosive learning in humans. *Scientific Reports* **8**, 2259 (2018).

[268] Holland, P. W., Laskey, K. B. & Leinhardt, S. Stochastic blockmodels: First steps. *Social networks* **5**, 109–137 (1983).

[269] Snijders, T. A. & Nowicki, K. Estimation and prediction for stochastic blockmodels for graphs with latent block structure. *Journal of Classification* **14**, 75–100 (1997).

[270] Nowicki, K. & Snijders, T. A. B. Estimation and prediction for stochastic blockstructures. *Journal of the American Statistical Association* **96**, 1077–1087 (2001).

[271] Airoldi, E. M., Blei, D. M., Fienberg, S. E. & Xing, E. P. Mixed membership stochastic blockmodels. *Journal of Machine Learning Research* **9**, 1981–2014 (2008).

[272] Goldenberg, A., Zheng, A. X., Fienberg, S. E., Airoldi, E. M. *et al.* A survey of statistical network models. *Foundations and Trends® in Machine Learning* **2**, 129–233 (2010).

[273] Qin, T. & Rohe, K. Regularized spectral clustering under the degree-corrected stochastic blockmodel. In *Advances in Neural Information Processing Systems*, 3120–3128 (2013).

[274] Anandkumar, A., Ge, R., Hsu, D. & Kakade, S. M. A tensor approach to learning mixed membership community models. *The Journal of Machine Learning Research* **15**, 2239–2312 (2014).

[275] Esquivel, A. V. & Rosvall, M. Compression of flow can reveal overlapping-module organization in networks. *Physical Review X* **1**, 021025 (2011).
[276] Fortunato, S. Community detection in graphs. *Physics reports* **486**, 75–174 (2010).
[277] Bazzi, M. *et al.* Community detection in temporal multilayer networks, with an application to correlation networks. *Multiscale Modeling and Simulation: A SIAM Interdisciplinary Journal* **14**, 1–41 (2016).
[278] Bazzi, M. *et al.* Community detection in temporal multilayer networks, with an application to correlation networks. *Multiscale Modeling & Simulation* **14**, 1–41 (2016).
[279] Gauvin, L., Panisson, A. & Cattuto, C. Detecting the community structure and activity patterns of temporal networks: a non-negative tensor factorization approach. *PloS one* **9**, e86028 (2014).
[280] Valles-Catala, T., Massucci, F. A., Guimera, R. & Sales-Pardo, M. Multilayer stochastic block models reveal the multilayer structure of complex networks. *Physical Review X* **6**, 011036 (2016).
[281] Taylor, D., Shai, S., Stanley, N. & Mucha, P. J. Enhanced Detectability of Community Structure in Multilayer Networks through Layer Aggregation. *Physical Review Letters* **116**, 228301 (2016).
[282] Taylor, D., Caceres, R. S. & Mucha, P. J. Super-Resolution Community Detection for Layer-Aggregated Multilayer Networks. *Physical Review X* **7**, 031056 (2017).
[283] Rosvall, M., Esquivel, A. V., Lancichinetti, A., West, J. D. & Lambiotte, R. Memory in network flows and its effects on spreading dynamics and community detection. *Nature Communications* **5**, 4630 (2014).
[284] Salnikov, V., Schaub, M. T. & Lambiotte, R. Using higher-order markov models to reveal flow-based communities in networks. *Scientific reports* **6**, 23194 (2016).
[285] Lambiotte, R., Rosvall, M. & Scholtes, I. Understanding complex systems: From networks to optimal higher-order models. *preprint arXiv:1806.05977* (2018).
[286] Aicher, C., Jacobs, A. Z. & Clauset, A. Learning latent block structure in weighted networks. *Journal of Complex Networks* **3**, 221–248 (2015).
[287] Szell, M., Lambiotte, R. & Thurner, S. Multirelational organization of large-scale social networks in an online world. *PNAS* **107**, 13636–13641 (2010).
[288] Pomeroy, C., Dasandi, N. & Mikhaylov, S. J. Multiplex communities and the emergence of international conflict. *PloS one* **14**, e0223040 (2019).
[289] Aleta, A., Tuninetti, M., Paolotti, D., Moreno, Y. & Starnini, M. Link prediction in multiplex networks via triadic closure. *Physical Review Research* **2**, 042029 (2020).
[290] Gallotti, R. & Barthelemy, M. Anatomy and efficiency of urban multimodal mobility. *Scientific Reports* **4**, 1–9 (2014).
[291] Didier, G., Brun, C. & Baudot, A. Identifying communities from multiplex biological networks. *PeerJ* **3**, e1525 (2015).
[292] Halu, A., De Domenico, M., Arenas, A. & Sharma, A. The multiplex network of human diseases. *NPJ systems biology and applications* **5**, 1–12 (2019).
[293] Choobdar, S. *et al.* Assessment of network module identification across complex diseases. *Nature methods* **16**, 843–852 (2019).
[294] Valdeolivas, A. *et al.* Random walk with restart on multiplex and heterogeneous biological networks. *Bioinformatics* **35**, 497–505 (2019).
[295] De Domenico, M., Sasai, S. & Arenas, A. Mapping multiplex hubs in human functional brain networks. *Frontiers in neuroscience* **10**, 326 (2016).
[296] Yu, M. *et al.* Selective impairment of hippocampus and posterior hub areas in alzheimer?s disease: an meg-based multiplex network study. *Brain* **140**, 1466–1485 (2017).

[297] Battiston, F., Nicosia, V., Chavez, M. & Latora, V. Multilayer motif analysis of brain networks. *Chaos: An Interdisciplinary Journal of Nonlinear Science* **27**, 047404 (2017).

[298] Frolov, N. S. *et al.* Macroscopic chimeralike behavior in a multiplex network. *Physical Review E* **98**, 022320 (2018).

[299] Buldú, J. M. & Porter, M. A. Frequency-based brain networks: From a multiplex framework to a full multilayer description. *Network Neuroscience* **2**, 418–441 (2018).

[300] Lim, S., Radicchi, F., van den Heuvel, M. P. & Sporns, O. Discordant attributes of structural and functional brain connectivity in a two-layer multiplex network. *Scientific reports* **9**, 1–13 (2019).

[301] De Domenico, M., Nicosia, V., Arenas, A. & Latora, V. Structural reducibility of multilayer networks. *Nature Communications* **6**, 6864 (2015).

[302] Passerini, F. & Severini, S. Quantifying complexity in networks: the von neumann entropy. *International Journal of Agent Technologies and Systems* **1**, 58–67 (2009).

[303] De Domenico, M. & Biamonte, J. Spectral entropies as information-theoretic tools for complex network comparison. *Physical Review X* **6**, 041062 (2016).

[304] Boguna, M., Krioukov, D. & Claffy, K. C. Navigability of complex networks. *Nature Physics* **5**, 74 (2009).

[305] Muscoloni, A. & Cannistraci, C. V. Navigability evaluation of complex networks by greedy routing efficiency. *Proceedings of the National Academy of Sciences* **116**, 1468–1469 (2019).

[306] Kamada, T., Kawai, S. *et al.* An algorithm for drawing general undirected graphs. *Information processing letters* **31**, 7–15 (1989).

[307] Fruchterman, T. M. & Reingold, E. M. Graph drawing by force-directed placement. *Software: Practice and experience* **21**, 1129–1164 (1991).

[308] Martin, S., Brown, W. M. & Wylie, B. N. Drl: Distributed recursive (graph) layout. Tech. Rep., Sandia National Laboratories (2007).

Glossary

Adjacency matrix : mathematical representation of a complex network, encoding information about which node is connected to which other node and with which intensity. Asymmetric matrices represent directed networks, while symmetric matrices represent undirected ones. Binary matrices encode unweighted networks, while non-binary matrices represent weighted networks.

Aggregate network : monoplex network obtained by aggregating (e.g., pairwise sum or other binary matrix operations) the layers of a multilayer network.

Annular visualization : a special type of visualization allowing one to highlight correlation patterns between measures obtained from multilayer networks. For instance, it allows one to compare the values of single-layer node centrality across multiple layers or to compare multiple versatility measures simultaneously.

Assortative mixing : see Assortativity.

Assortativity : tendency of nodes to be connected to other nodes of similar kind.

Authority centrality : see HITS centrality.

Average path length : sample average of the length of all shortest paths in a network.

Barabasi-Albert model : network growth model characterized by preferential attachment linking that yields a power-law degree distribution.

Betweenness centrality : centrality descriptor proportional to the number of shortest paths that cross each node.

Brain networks : networks where nodes are regions of interest within the human brain and edges indicate the presence of physical or functional (e.g., based on statistical correlations) relationships among them.

Centrality : score usually assigned to each node of a network to characterize their importance with respect to a set of criteria encoded by algorithms (e.g., number of links, attraction of information flow, etc.)

Circuit : sequence of nodes and edges traversed by a walker. It is a closed trail with no repeated edges.

Closeness centrality : centrality descriptor defined as the inverse of the sum of geodesic (i.e., shortest-path) distances between one node and the other nodes in a network.

Cluster : a group of nodes within a community (see Group) or within a connected component (see Connected component).

Clustering : referring to triadic closure, it provides a local or global measure of the tendency of nodes to form triangles.

M. De Domenico, *Multilayer Networks: Analysis and Visualization*, https://doi.org/10.1007/978-3-030-75718-2

Codeword : bit string used to encode a sequence of symbols.

Community : see Group.

Complex network : set of units (nodes) and relationships (edges) organized into non-trivial connectivity patterns.

Configuration model : random network model used to build uncorrelated networks with a given degree distributions.

Connected component : subsets of the nodes that are reachable by undirected paths.

Connectome : networks where nodes are neural units (e.g., neurons, brain areas) and edges indicate their physical or functional relationships. See also Brain networks.

Coreness centrality : centrality descriptor assigning a score k to a node if it belongs to the k-core of the network and it does not belong to the $(\ell+1)$-core. See also k-core decomposition.

Coupled layers : distinct networks which are connected together by structural or functional relationships.

Coverage : in a stochastic process on the top of a complex network, it is defined as the fraction of nodes visited by a random walker at least once within a certain amount of time.

Critical point : value of the control parameter of a system at which a phase transition occurs.

Cycle : sequence of nodes and edges traversed by a walker. It is a closed path with no repeated nodes and edges.

Degree centrality : centrality descriptor counting the number of edges (incoming, outgoing, inter-layer, intra-layer) incident on one node.

Degree-degree correlation : Influence that the degree of a node has on the degree of its neighbors.

Density matrix : in the statistical field theory of information dynamics on networks, a matrix which encodes the state of a network as obtained from superposing information streams weighted by their activation probabilities.

Description length : given a data set and a hypothesis to describe it (e.g., a mathematical model), the description length is defined as the sum of two terms, one encoding the number of bits required to describe the data according to the hypothesis and a second one encoding the number of bits required to describe the hypothesis.

Diameter : the longest shortest path in a network.

Disassortativity : the opposite of Assortativity.

Diffusion : a special case of dynamics on the top of a network, where (stochastic) rules are used to distribute information (e.g., water, a meme, etc) in the neighborhood of a node.

Dimensionality reduction : see Reducibility.

Dismantling : Process consisting of the deliberate removal of nodes and/or links of a network according to a specific mechanistic rule (e.g., according to scores assigned to nodes).

Dynamics : used to indicate a dynamical process defined on the top of the network (e.g., a random walk, synchronization, consensus, etc.) or the rules governing the growth or shrink of the network.

Ecological networks : networks where nodes are ecological units (e.g., plants, pollinators, predators, preys, etc.) and edges indicate their relationships (cooperation, predation, mutualism, parasitism, etc.).

Edge : connection encoding interaction or any type of relationship between two nodes in a complex network.

Edge-colored multigraph : multilayer network where layers are not interconnected with each other.

Eigenvector centrality : iterative procedure to assign a score to each node in a network while accounting for the importance of its neighbors, the neighbors of neighbors, so forth and so on. It is estimated as the dominant eigenvector of a governing matrix: in the simplest case, such a matrix is the adjacency one. It is not well defined for directed networks.

Ensemble, network : see Random network model(s).

Emergence : in complex systems, emergent phenomena which are observed at higher scales than microscopic one, and can not be simply deduced from the full knowledge of nodes only.

Erdös-Rényi model : random network model constructed by connecting every pair of nodes independently with the same probability.

Exponential Random Graph model : random network model constructed by maximizing a functional (specifically, the Gibbs entropy) according to a set of constraints (e.g., the observed average degree, the degree sequence, etc) which are encoded within a Hamiltonian function.

Flattening : see Matricization.

Functional network : a complex network where edges encode measures of intensity or relationships, based on statistical correlations or similarities, between any pair of nodes. See Structural network for comparison.

Giant connected component : order parameter of the percolation phase transition. It is exactly 0 in the non-percolating phase and becomes proportional to the system size in the percolating phase.

Group : sub-set of nodes organized to exhibit a denser connectivity (i.e., number of links, information flow, so forth and so on) among themselves than with other nodes in a network. Also known as community, cluster, module, the organization of nodes into groups defines the mesoscale of the system, which is usually identified by means of community detection methods.

HITS centrality : a type of eigenvector centrality, consisting of two descriptors, hub and authority. The governing matrix is given by the product of the adjacency matrix by its transpose for the hub score, and by the product of the transpose of the adjacency matrix by the adjacency matrix for the authority score.

Hub centrality : see HITS centrality.

Information : amount of uncertainty (or surprise) of a variable outcome. Information is measured in bits, where an information of 1 bit reduces event uncertainty by half.

Information-theoretic : based on information theory.

Information stream : in the statistical field theory of information dynamics on networks, a matrix obtained from the eigen-decomposition of a propagator, e.g. the diffusion propagator, which drives the flow of information through the network.

Interactome : see Protein-protein interaction networks.

Interconnected network : see Coupled layers.

Interdependent network : multilayer network where layers are interconnected with each other, and each physical node is defined in only one layer, thus coinciding with its state node.

Inter-layer : defined across the layers of a multilayer network, e.g. to indicate correlations or operations involving at least two distinct layers.

Intra-layer : defined within a layer of a multilayer network, e.g. to indicate correlations or operations involving only one layer.

Jensen-Shannon divergence : symmetric information-theoretic measure used to quantify the difference between two probability distributions.

Jensen-Shannon distance : Square root of the Jensen-Shannon divergence allowing one to define an information-theoretic metric.

Katz centrality : a type of eigenvector centrality which overcomes the limitations of eigenvector centrality.

k-core : maximal subgraph consisting of nodes with degree equal to or larger than k.

k-core decomposition : procedure to find all k-cores in a network.

Kullback-Leibler divergence : asymmetric information-theoretic measure used to quantify the difference between two probability distributions

Information entropy : see Shannon entropy.

Information entropy, network : see Von Neumann entropy.

Lancichinetti-Fortunato-Radicchi model : random network model that produces synthetic networks with communities whose degree distribution and community size distribution are both power law.

Laplacian matrix : known also as discrete Laplacian or Kirchhoff matrix, it provides another matrix representation of a network than adjacency matrix. There are many Laplacian matrices, which differ in the way they are normalized and the type of dynamics they govern by means of a master equation. In the simplest case, the Laplacian matrix governs simple diffusion through a network: its diagonal entries correspond to degrees (or strengths) of nodes, while its off-diagonal entries equal -1 (or the negative weight) if two nodes are adjacent and 0 otherwise.

Largest connected component : the largest among the connected components in a network. In finite-size simulations, it is used to approximate the giant connected component.

Largest intersection component : the largest cluster in which nodes are connected across all layers simultaneously.

Largest viable component : the largest cluster in which nodes are connected by a path in each layer simultaneously.

Layer : in a multilayer network, it is a sub-system represented by a network which encodes a specific type of interactions among nodes.

Layer-layer correlation : topological correlation occurring between at least two networks within a multilayer system. Usually, it is measured through statistical correlation of descriptors (e.g., degree, clustering) measured from the corresponding networks. It can also refer to overlapping topological units, such as nodes and edges (see Overlapping).

Link : see Edge.

Markovian process : stochastic process whose next state depends only on the current system state.

Matricization : algebraic operation which transforms a tensor into another tensor by changing its shape and dimension, without loss of information.

Mean path length : see Average path length.

Mesoscale : organization of nodes into groups.

Metabolic networks : networks where nodes are metabolites and edges indicate their relationships (physical, chemical, etc.).

Metabolome : see Metabolic networks.

Minimum description length (MDL) : a principle which formalizes Occam's razor to describe data by using the shortest possible description in terms of bit strings. It is used as a model selection principle: the shortest description corresponds to the most parsimonious description of the data. See also Description length.

Modularity : a measure of the organization of a network in modules (or groups). It quantifies the tendency of modules to be more dense than expected under the assumption that edges were randomly distributed according to the network?s configuration model. See also Modularity and Configuration model.

Modularity maximization : optimization procedure, based on modularity, used in community detection to identify groups or communities in a network. See also Modularity.

Module : see Group.

Molecular networks : networks where nodes are molecules (e.g., genes, proteins, metabolites, etc.) and edges indicate their relationships (physical, chemical, etc.).

Monoplex network : see Complex network.

Motifs : subgraphs, usually consisting of a few nodes, which significantly recur within a network.

Multiplex network : multilayer network where layers are interconnected with each other in such a way that each node in a layer is linked to its own replicas in other layers.

Multiplexity : feature of a system where nodes exhibit multiple types of pairwise interactions. Also referred to single nodes (e.g., node multiplexity) and edges (e.g., edge multiplexity) in presence of overlapping nodes and edges, respectively.

Multilayer network : a network consisting of networks coupled together by structural or functional relationships. Also known as a system of systems. Special types of multilayer networks are Edge-Colored multigraphs, Multiplex networks and Interdependent networks.

Multilayer adjacency tensor : a self-consistent mathematical representation of a multilayer network, where entries encode connections between nodes within and across layers.

Multilayer Laplacian tensor : Laplacian tensor corresponding to a multilayer adjacency tensor.

Multimap : information-theoretic method to identify communities in a multilayer network. It is based on the multilayer generalization of the map equation, which in turn is based on multilayer random walks and the compression of their trajectories. The optimization procedure is based on the minimum description length principle.

Mutual information : symmetric information-theoretic measure to quantify the similarity between two probability distributions. It can be obtained as a special case of the Kullback-Leibler divergence.

Navigability : feature of a complex network describing and quantifying how easy it is to explore its structure in terms of a specific dynamics (e.g., a random walk).

Network : Mathematical object composed by a set of nodes and a set of edges, which are ordered pairs of nodes.

Network of layers : connectivity pattern (i.e., a complex network) describing how layers within a multilayer network are coupled together.

Node : fundamental unit of a complex network, used for the abstract representation of an entity (e.g., a protein, a neuron, an individual, a geographic area, a word, etc.)

Nonlinear dynamics : dynamical process describing a system in which a change of the output is not proportional to a change of the input.

Non-Markovian process : a stochastic process which is not Markovian.

Normalized information loss : information-theoretic measure used to quantify the goodness of a network partition.

Null network model : network model, often based on stochastic processes, used as a null model for the statistical analysis of a complex network.

Overlapping, node : a node with more than one replica/state-node in a multilayer network.

Overlapping, edge : an edge connecting the same pair of nodes across at least two layers in a multilayer network.

Page-rank centrality : a type of eigenvector centrality which overcomes the limitations of eigenvector and Katz centralities. It can also be interpreted as the steady state of a random walk (i.e., a Markov process) governed by a special transition matrix known as Google matrix.

Partition : in the context of community detection (see Group), it refers to the assignment of each node to (at least) one community. The set of communities defines a partition.

Partition function : in the statistical field theory of information dynamics on networks, it corresponds to the summation of the stream sizes and it is used to normalize the density matrix.

Path : sequence of nodes and edges traversed by a walker. It is a trail with no repeated nodes and edges.

Percolation : Process that consists of removing a fraction of nodes (site percolation) and/or links (bond percolation) following a given criterion, to later compute statistical and geometrical properties of the remaining subsystem.

Phase transition : change in a system from one state to another characterized by a different microscopic organization. The order parameter characterizes the system organization and changes with infinitesimal variations of some control parameters.

Physical node : set of state nodes (or replicas) defining the identity of a node in a multilayer network. See also Node.

Power law : Relation between two quantities where one varies as a power of the other. The main properties are the scale invariance and the divergence of the moments depending on the exponent of the relation.

Protein-protein interaction networks : networks where nodes are protein and edges indicate their relationships (physical, chemical, genetic, statistical, etc.).

Quantum Jensen-Shannon divergence : symmetric information-theoretic measure used to quantify the difference between two density matrices.

Quantum Jensen-Shannon distance : Square root of the Quantum Jensen-Shannon divergence allowing one to define an information-theoretic metric.

Quantum Kullback-Leibler divergence : asymmetric information-theoretic measure used to quantify the difference between two density matrices.

Random network model(s) : a class of models used to reproduce one or more features observed in empirical complex networks. They are based on mechanistic assumptions where connectivity is built by means of stochastic rules (e.g., the rules for growth or for wiring two nodes) defining network ensembles. Popular examples of random network models are the Erdös-Rényi model, the Barabasi-Albert model, the Stochastic Block Model (SBM), the Lancichinetti-Fortunato-Radicchi (LFR) model, the Configuration Model (CM), as well as the family of Exponential Random Graph Models (ERGM).

Reducibility : procedure devoted to coarse-grain a multilayer network to reduce its size and complexity. It usually depends on a cost function, quantifying the loss of information during the coarse-graining procedure, and can be either based on structural or functional measures.

Replica, node : see State node.

Random failure : Unpredictable failure occurring in a network, usually modeled as randomly uniform removal of nodes and/or links.

Random walk : a walk where the choice of edges to use (or the choice of nodes to jump into) is made stochastically, according to some transition rules.

Self-loop : an edge which starts and ends in the same node of a network.

Shannon entropy : average amount of information or surprise a receiver has with respect to the possible outcomes of a message sent, through a communication channel, by a sender. Often, it is used to quantify the level of uncertainty about the value of a stochastic variable with known probability distribution: to this extent, it can be understood as a measure of flatness, i.e., how such a distribution is uniform.

Shortest path : among all paths connecting two nodes, it is the one with shortest length.

Similarity : a measure of correlation between two vectors (or matrices).

SNXI decomposition, dynamical : special decomposition of a multilayer dynamics in terms of four different contributions, each one corresponding to a specific set of dynamical effects: self-relationships, endogenous, exogenous and intertwining.

SNXI decomposition, structural : special decomposition of a multilayer adjacency tensor in terms of the contributions of four different tensors, each one corresponding to a specific set of structural relationships: self-relationships, endogenous, exogenous and intertwining interactions.

Social networks : networks where nodes are individuals with edges indicating the presence of a social relationship (e.g., trust, friendship, business, etc) among them.

Socio-technical networks : networks where nodes are individuals and non-human units (e.g., machines), or individuals within a technological context (e.g., online platforms), with edges indicating the presence of a relationship among them.

Socio-ecological networks : networks where nodes are individuals and ecological units (e.g., species), or individuals within an ecological context, with edges indicating the presence of a relationship among them.

Spectral entropy, network : see Von Neumann entropy.

State node : node defined within a specific layer of a multilayer network. Also known as node replica.

Statistical physics : area of physics focused on the study of large ensembles of interacting entities and whose scope is to explain system-wide properties based on the local interactions.

Stochastic block model : random network model that produces synthetic networks with mesoscale structure (e.g., communities, core-periphery, etc) governed by a specific connectivity matrix fixing the probability that two nodes are linked within a block (i.e., a group) or across blocks.

Strength centrality : centrality descriptor summing the weights of edges (incoming, outgoing, inter-layer, intra-layer) incident on one node. For undirected networks, it coincides with the degree centrality.

Strongly connected component : in a directed network, subsets of the nodes that are mutually reachable by directed paths.

Structure : referred to the topology of a complex network.

Structural network : a complex network where edges encode measures of intensity or relationships, not based on statistical correlations or similarities, between any pair of nodes. Functional network for comparison.

Supra-adjacency matrix : a possible mathematical representation of a multilayer network, where entries encode connections between nodes within and across layers.

Supra-Laplacian matrix : Laplacian matrix corresponding to a supra-adjacency matrix.

System : set of units characterized by interactions and/or other types of relationships. Usually used as a synonym of complex network, although not all complex systems admit a complex network representation.

System of systems : see Multilayer network.

Targeted attacks : Removal of nodes and/or links based on some a priori knowledge about the network.

Temporal network : a complex network where nodes and links can change over time.

Time-varying network : see Temporal network.

Topology : see Structure.

Tensor : a fundamental mathematical object able to encode multilinear relationships between units defined in a vector space.

Trail : sequence of nodes and edges traversed by a walker. It is an open walk with no repeated edges.

Transitivity : see Clustering.

Transportation networks : networks where nodes are geographic points (e.g., stations, airports) and edges indicate the presence of a route among them.

Triadic closure : see Clustering.

Urban networks : networks where nodes are geographic areas within a city and edges indicate their physical connections (e.g., roads).

Versatility : generalization of the concept of node centrality to the realm of multilayer networks.

Vertex : see Node.

Von Neumann entropy : mixedness of information streams obtained from the network density matrix. It is the generalization of Shannon entropy to the case of quantum systems and complex networks.

Walk : sequence of nodes and edges traversed by a walker. Nodes and edges can be repeated. If a walk starts and ends at the same node, it is called a closed walk, otherwise it is an open walk.

Watts-Strogatz model : Network model characterized, simultaneously, by a high clustering and a low average path length, obtained by rewiring a fraction of the links of one-dimensional ring lattices.

Weakly connected component : in a directed network, subsets of the nodes that are reachable by undirected paths.

Printed in the United States
by Baker & Taylor Publisher Services